THE HISTORY
OF SCIENCE AND TECHNOLOGY
IN THE UNITED STATES

BIBLIOGRAPHIES ON THE HISTORY
OF SCIENCE AND TECHNOLOGY
(VOL. 17)

GARLAND REFERENCE LIBRARY
OF THE HUMANITIES
(VOL. 815)

Bibliographies on the History of Science and Technology

Editors

Robert Multhauf, Smithsonian Institution, Washington, D.C.
Ellen Wells, Smithsonian Institution, Washington, D.C.

THE HISTORY
OF SCIENCE AND TECHNOLOGY
IN THE UNITED STATES
A Critical and Selective Bibliography
Volume II

Marc Rothenberg

GARLAND PUBLISHING, INC. • NEW YORK & LONDON
1993

© 1993 Marc Rothenberg

Library of Congress Cataloging-in-Publication Data
(Revised for vol. 2)

Rothenberg, Marc, 1949–
  The history of science and technology in the United States.
  (Bibliographies on the history of science and technology ; v. 2,
17   ) (Garland reference library of the humanities ; vol. 308, 815)
  Includes indexes.
  1. Science—United States—History—Bibliography. 2. Technol-
ogy—United States—History—Bibliography.
Z7405.H6R67   1982   [Q125]   016.50973   81–43355
ISBN 0–8240–8349–0 (v. 2 : alk. paper)

Printed on acid-free, 250-year-life paper
Manufactured in the United States of America

To Ivy with love

GENERAL INTRODUCTION

This bibliography is one of a series designed to guide the reader into the history of science and technology. Anyone interested in any of the components of this vast subject area is part of our intended audience, not only the student, but also the scientist interested in the history of his own field (or faced with the necessity of writing an "historical introduction") and the historian, amateur or professional. The latter will not find the bibliographies "exhaustive," although in some fields he may find them the only existing bibliographies. He will in any case not find one of those endless lists in which the important is lumped with the trivial, but rather a "critical" bibliography, largely annotated, and indexed to lead the reader quickly to the most important (or only existing) literature.

Inasmuch as everyone treasures bibliographies it is surprising how few there are in this field. Justly treasured are George Sarton's *Guide to the History of Science* (Waltham, Mass., 1952; 316 pp.), Eugene S. Ferguson's *Bibliography of the History of Technology* (Cambridge, Mass., 1968; 347 pp.), François Russo's *Histoire des Sciences et des Techniques, Bibliographie* (Paris, 2nd ed., 1969; 214 pp.), and Magda Witrow's *ISIS Cumulative Bibliography. A bibliography of the history of science* (London, 1971–1985; Boston 1989– ; 9 vols. as of 1989). But all are limited, even the latter, by the virtual impossibility of doing justice to any particular field in a bibliography of limited size and almost unlimited subject matter.

For various reasons, mostly bad, the average scholar prefers adding to the literature, rather than sorting it out. The editors are indebted to the scholars represented in this series for their willingness to expend the time and effort required to pursue the latter objective. Our aim has been to establish a general framework which will give some uniformity to the series, but otherwise to leave the format and contents to the author/compiler. We have urged that introductions be used for essays on "the state of the field," and that selectivity be exercised to limit the length of each volume to the economically practical.

Since the historical literature ranges from very large (e.g., medicine) to very small (chemical technology), some bibliographies will be limited to the most important writings while others will include modest "contributions" and even primary sources. The problem is to give useful guidance into a particular field—or subfield—and its solution is largely left to the author/compiler.

In general, topical volumes (e.g., chemistry) will deal with the subject since about 1700, leaving earlier literature to the area of chronological volumes (e.g., medieval science); but here, too, the volumes will vary according to the judgment of the author. The topics are international, with a few exceptions, but the literature covered depends, of course, on the linguistic equipment of the author and his access to "exotic" literatures.

Robert Multhauf
Ellen Wells
Smithsonian Institution
Washington, D.C.

CONTENTS

PREFACE

This volume is a follow up to *The History of Science and Technology in the United States: A Critical and Selective Bibliography* (Garland, 1982). Like its predecessor, it is a guide to the secondary literature in the history of science and technology, excluding medicine. It has been compiled to assist newcomers--whether graduate students or experienced scholars from other segments of the history of science and technology, or American history--by providing a general orientation to the literature.

The first volume considered earlier publications. The current volume contains only publications that originally appeared between 1980 and 1987 and includes books, articles, dissertations, and essay reviews. All publications listed are in English, which reflects the fact that almost all historians of American science and technology publish in that language.

This bibliography is a distillation of more than twelve hundred works. To be included in the bibliography, a publication had to meet two criteria. First, it had to be useful to a non-specialist. The broader the potential audience and the larger the questions asked, the higher the probability that a publication would be included. Second, the publication had to be either of fundamental importance for an understanding of an aspect of the history of American science and technology, or illustrative of the great range of methods and sources used by historians of these fields. The use of these two criteria resulted in the exclusion of many highly technical, narrowly cast articles in the history of the various disciplines, despite their obvious importance in understanding the development of specific disciplines.

Dissertations were treated in a special way. They were included only if they were not adequately represented in the pre-1988 published literature. In those cases in which I am aware of a post-1987 publication based on a dissertation, I have included the citation of the published version at the end of the annotation. However, the annotation describes the dissertation, not the publication.

As was the case for its predecessor volume, this bibliography does not include works in the sociology of science and science policy studies, even if they used a historical approach.

Entries are categorized by topic. Within categories, the entries are listed alphabetically by author. Many entries could be placed logically in more than one category; however, with one exception, I have placed entries under what I judge to be the single most useful category. The subject index will guide the user to all entries of interest.

Annotations are descriptive and often evaluative. Generally, annotation length reflects the complexity of the entry and the need to summarize it accurately. There is no correlation between the quality of a work and the length of its annotation.

References to the predecessor volume of this bibliography are given in the annotations in the form I:xxx, where xxx refers to an entry number.

Numerous individuals assisted in the creation of this bibliography. I want to thank the interlibrary loan staff of the Smithsonian Institution Libraries. Their assistance was fundamental to the success of this volume. Many individual scholars sent me reprints or nominated entries for the bibliography. They saved me the embarrassment of missing important contributions to the literature. My wife, Ivy Baer, corrected my grammar. Finally, I want to thank Ivy and my daughters, Sara, and Leslie, for their tolerance of this project. It took me away from them on many weekends and evenings.

INTRODUCTION

When I prepared Volume I of this bibliography, I did not explicitly define the subject matter, "science and technology in the United States." It seemed unnecessary at that time, since I felt that there was an implicit consensus about the boundaries of the field. Furthermore, the term seemed interchangeable with the term "American science and technology."

Subsequently, I learned that this was not so, particularly in the history of science. Not only do some historians feel that "science in the United States" and "American science" are not synonymous, but semantic differences may signal significant differences in philosophy, historiographic orientation, or professional allegiance. For some scholars (and this must be discerned from their writings or learned from oral discourse; historians are reluctant to provide explicit, written definitions), the term "in the United States" is used when signifying that an activity is occurring within particular geographical boundaries. It may also be used as an indication that the historiographic approach is intellectual history or the history of ideas. In contrast, the term "American science" is used in situations in which historians want to argue about the existence of unique American characteristics to the scientific practice in this country, or to indicate that the approach taken fits under the rubrics of cultural or social history. Those who speak of "science in the United States" seem to identity themselves primarily as historians of disciplines, while users of the term "American science" are more likely to label themselves as Americanists, and have an alliance with, if not an allegiance to, the field of American history. The former focus on the science, while the latter look at the interaction between scientific activities and the larger cultural context, whether institutional, intellectual, educational, or political, which is identified with this particular nation-state. (Historians of American technology do not seem to have these differences, at least not in public.)

My definition of the history of science and technology in the United States has five elements. The first is that I use the terms "American science," "American technology" and "science and technology in the United States" interchangeably, depending upon stylistic considerations, despite the technical impreciseness this represents. Second, the reference is to both Euroamerican and Native American activities, although I acknowledge that in reality little is available on the latter topic. Third, for the Colonial Era, the geographical boundaries include both the British and the Spanish colonies north of modern Mexico. Fourth, for the period after 1788, scientific or technological activity is labelled American if the participants were either citizens or permanent residents of the United States, or conducted the activity within the confines of an institution situated within that country, or had this activity sponsored by such an institution. According to this criteria, research on Americans studying in Germany and Germans studying in the United States might both be American topics; so might the domestic and international activities of the Du Pont Corporation.

* * *

The field of the history of science in the United States is active, almost to the point of overwhelming anyone attempting to keep pace with developments. This bibliography is one measure of that activity, which, in the years subsequent to those covered here, appears to be increasing.

One of the great strengths of the field is its institutional breadth. For example, during 1989 and 1990, fifty-two dissertations on topics in the history of American science were accepted by graduate programs in the United States and Canada. Thirty-seven American and two Canadian institutions granted these degrees. No institution was responsible for more than three graduates. In most cases, the degrees were granted by history departments, not independent history of science departments.

The usual explanation given for increasing interest in American topics is concern with twentieth-century, including contemporary, science where the United States looms as the major player. Statistics seem to support the idea that many Americanists are interested in the twentieth century. Of the forty-two papers on American topics on the program of the 1992 meeting of the History of Science Society, thirty-four deal with twentieth-century issues, either in whole or in part. Thirty-five of the fifty-two dissertations mentioned above dealt with twentieth-century subjects. Although interest in the nineteenth and eighteenth centuries has not disappeared, the history of American science is dominated by students of the more recent past.

Historians of American technology also are very concerned with the twentieth century. Twenty-four of thirty-eight papers given at either the 1990 or 1991 meetings of the Society for the History of Technology dealt in whole or in part with this century, despite a deliberate effort by the Society to be more international in both membership and intellectual thrust.

* * *

There are a few points which I made in the "Introduction" to Volume I which deserve a follow-up. The first was the lack of scholarly biographies in the history of American science. There has been improvement during the last decade, but many significant figures still await a biographer. Thanks to the *American National Biography*, to be published by Oxford University Press sometime in the future, almost every American scientist of some reputation will have at least one thousand words written about him or her. Will the research that went into these short biographical sketches spark interest among historians to produce book-length studies?

In my previous introduction I deplored the quality of the work in the history of the social sciences. I criticized it for being "the home of the most practitioner history," and described it as being isolated from other work in the history of American science. During the intervening years the quality of the scholarship has improved considerably. Anthropology continues to be the leading area, but both sociology and the behavioral sciences have closed the gap.

Increasing quality also has been a characteristic of the history of the National Aeronautics and Space Administration, which in 1982 I felt was a near disaster. Independent acknowledgement of this improvement may be found in item 202 of this bibliography, a fine defense of NASA history by A. Hunter Dupree.

* * *

Readers who compare the topical categories used in Volume I and Volume II will find some differences, especially in the categories gathered under the rubric of "Special Themes." In some cases, there has been only a change in the title; e.g., "Funding" became "Patronage." In other cases, there has been a change in philosophy. An example of this is the treatment of entries on women. In Volume I, they were all gathered together in the category "Science, Technology, and Women." To avoid the unnecessary segregation of works about women, Volume II integrates such entries within appropriate scientific and technological categories. The index allows the user to recover them as a group. I also deleted "Science, Technology, and Government" as a category under "Special Themes" and made it a separate chapter.

Some changes in the bibliography, however, reflect what I believe to be changes in the field. Two categories prominent in Volume I-- "Learned Societies" and "Professionalization"--do not appear in

Volume II. When the focus of American history of science was on the eighteenth and nineteenth centuries, these were significant categories. For historians of twentieth-century science, neither the American Philosophical Society nor the American Academy of Arts and Sciences holds much interest. Neither do these historians debate the differences between amateur and professional scientists. Added for Volume II was "Art and Literature." Historians of science and technology and historians of art and literature have become increasing aware of the impact of their respective areas of study upon each other.

Entries in the history of technology constituted approximately 30 percent of Volume I. For Volume II, the figure is 40 percent. This difference may be more an artifact of my increasing awareness of the literature in the history of technology than a reflection of the changing relative strengths of the fields.

THE HISTORY
OF SCIENCE AND TECHNOLOGY
IN THE UNITED STATES

CHAPTER I: BIBLIOGRAPHIES AND GENERAL STUDIES

BIBLIOGRAPHIES

001. Beardsley, Edward H. "The History of American Science and Medicine." *Information Sources in the History of Science and Medicine.* Edited by Pietro Corsi and Paul Weindling. London: Butterworth Scientific, 1983, pp. 411-435.

Surveys the literature from about 1965 through 1979. Provides approximately one hundred bibliographic entries.

002. Cravens, Hamilton. "Science, Technology, and Medicine." *American Studies: An Annotated Bibliography.* Edited by Jack Salzman. New York: Cambridge University Press, 1986, pp. 1565-1668.

Provides 401 entries. Includes descriptive annotations. The coverage is uneven.

003. Elliott, Clark A. "Bibliographies, Reference Works, and Archives." *Osiris,* 2nd ser., 1 (1985): 295-310.

Reviews material from 1971 through 1984. Calls for better reference aids for documentary material.

004. Elliott, Clark A. "Some Recent Books." *History of Science in America: News and Views.*

Provides the most complete listing of books, monographs, and dissertations in the history of American science. Lacks annotations or indices. Appears once or twice a year.

005. Kidwell, Clara Sue. "Native Knowledge in the Americas." *Osiris*, 2nd ser., 1 (1985): 209-228.

Distinguishes between the European view that the lawful behavior of nature leads to prediction based on observation and the native belief that the forces of nature have a will and volition of their own. Contrasts the European use of experimentation with the native employment of ritual and ceremony. Provides guidance to the literature dealing with the ethnosciences.

006. Kohlstedt, Sally Gregory. "Institutional History." *Osiris*, 2nd ser., 1 (1985): 17-36.

Maintains that institutional history provides a convergent point for intellectual, social, and cultural history. Finds that studies of colonial and antebellum institutions are more comprehensive than for the later eras, where historians rely on case studies.

007. Kohlstedt, Sally Gregory, and Margaret W. Rossiter, editors. *Historical Writing on American Science: Perspectives and Prospects*. Baltimore: The Johns Hopkins University Press, 1986. 321 pp. Index.

Reprints *Osiris*, 2nd ser., 1 (1985). Includes items 003, 005, 006, 012, 162, 169, 177, 240, 264, 272, 286, 316, 388, 400.

008. Lowood, Henry, compiler. "Current Bibliography in the History of Technology." *Technology and Culture*. Indices.

Includes chronological and subject classifications. Identifies general references to American technology. Appears annually. This is a continuation of Item I:5.

009. Neu, John, editor. *Isis Cumulative Bibliography, 1976-1985. A Bibliography of the History of Science Formed from Isis Critical Bibliographies 101-110 Indexing Literature Published from 1975 through 1984*. Volume I: *Persons and Institutions*. Volume 2: *Subjects, Periods and Civilizations*. Boston: G. K. Hall: 1989. xiv+587 pp., xvi+911 pp. Indices.

Serves as an accumulation of item 010. Classifies North American topics as sub-entries of the subject index. Includes a cumulative book review listing. This is a continuation of Item I:7.

010. Neu, John, editor. *Isis Current Bibliography of the History of Science and Its Cultural Influences.* Index.

Includes discipline and chronological classifications. The index is limited to proper names and authors, making it difficult to scan for entries relating to the history of American science. Appears annually. This is a continuation of Item I:6.

011. Reingold, Nathan. "Clio as Physicist and Machinist." *Reviews in American History,* 10 (1982): 264-280.

Summarizes and critiques the literature of the history of American science and technology from 1971 through 1980. Describes most historians of American science as concentrating on local or case studies. Believes that historians of American technology have some status anxieties. Describes their work as a modernized, refined form of traditional economic history.

012. Roland, Alex. "Science and War." *Osiris,* 2nd ser., 1 (1985): 247-272.

Observes that there is an abundant amount of monographic literature, but argues that the histories of science and war have not been well integrated into surveys of American history. Offers an outline of what a synthesis of the existing literature might look like. Organizes the literature by topic within chronological periods. Defends official histories as better sources than academic historians have realized.

013. Roland, Alex. "Technology and War: A Bibliographic Essay." *Military Enterprise and Technological Change: Perspectives on the American Experience* (item 176), pp. 347-379.

Supplies a critical bibliography which is not limited to American history. Focuses on books.

014. Siegel, Patricia Joan and Kay Thomas Finley. *Women in the Scientific Search: An American Bio-Bibliography, 1724-1979.* Metuchen: The Scarecrow Press, 1985. xvii + 399 pp. Index.

Summarizes over fifteen hundred references to over two hundred and fifty scientists. Divides the women by discipline.

GENERAL SCIENCE

015. Beaver, Donald de B. *The American Scientific Community, 1800-1860: A Statistical-Historical Study.* New York: Arno, 1980. 379 pp. Appendices, Bibliography.

Reprints his 1966 dissertation (Item I:13).

016. Bruce, Robert V. *The Launching of Modern American Science, 1846-1876.* New York: Knopf, 1987. x+446 pp. Bibliography, Index.

Argues that the patterns and institutions in science established during the years 1846-1876 have endured into the late twentieth century. Provides a synthetic narration of the development of the community of professional scientists in the United States during this period. Focuses on the rise of institutions and the development of scientific careers, not experiments or the evolution of scientific ideas. Demonstrates the negative impact of the Civil War on American science. Treats technology in relation to science. This is the standard history for the period.

017. Elliott, Clark A. "Models of the American Scientist: A Look at Collective Biography." *Isis,* 73 (1982): 77-93.

Compiles data, for five periods, on five factors: place of birth, paternal occupation, education level, employment and scientific fields. Discusses a representative scientist for each period.

018. Freemon, Frank R. "American Colonial Scientists Who Published in the *Philosophical Transactions* of the Royal Society." *Notes and Records of the Royal Society of London,* 39 (1985): 191-206.

Evaluates forty-five articles published between 1753 and 1775. Concludes that the quality of the science was not inferior to a random sampling of European scientific papers in the same publication, but that the quantity was relatively insignificant.

019. Gillispie, Charles C. *The Professionalization of Science: France 1770-1830 Compared to the United States 1910-1970.* Kyoto: Doshisha University Press, 1983. 40 pp.

Contends that the distinctive element of American science was its scale: of men, money, and equipment.

020. Greene, John C. *American Science in the Age of Jefferson.* Ames: The Iowa State University Press, 1984. xiv+484 pp. Index.

Surveys the history of American science from 1780 to 1820. Considers this a period when progress was made through the establishment of institutional bases and the assimilation of European developments. Examines American science by location and discipline. Evaluates Thomas Jefferson's significance both as a promoter of American science and as a symbol of American respect for science. This is the standard history for this period.

021. Henderson, Janet Karen. "Four Nineteenth-Century Professional Women." Ed.D. dissertation, Rutgers University, 1982.

Studies the motivations, self-perceptions, and subjective responses of four pioneering professional women--Harriot Hunt, Elizabeth Blackwell, Maria Mitchell, and Ellen Swallow Richards. Argues that despite their breakthroughs, contemporary feminine stereotyping limited their productivity.

022. Hindle, Brooke. "Charles Willson Peale's Science and Technology." *Charles Willson Peale and His World.* Edgar P. Richardson, Brooke Hindle, and Lillian Miller. New York: Harry Abrams, 1982, pp. 106-169.

Presents Peale as an artisan who followed the apprenticeship mode of learning by observing and doing, whatever the endeavor. Credits his achievements to his ability to manipulate mental images.

023. Hindle, Brooke. "A Retrospective View of Science, Technology, and Material Culture in Early American History." *William and Mary Quarterly,* 3rd ser., 41 (1984): 422-435.

Reflects on his education and career. Credits the Institute of Early American History and Culture with significant support for the history of early American science and technology. Finds that interest in the history of pre-1820 American science peaked about 1955.

024. Hodes, Elizabeth. "Precedents for Social Responsibility among Sciences: The American Association of Scientific Workers and the Federation of American Scientists, 1938-1948." Ph.D. dissertation, University of California, Santa Barbara, 1982.

Contrasts the failure of the American Association of Scientific Workers to win broad support for its program of public education and social involvement by scientists with the success of the Federation of American Scientists in educating and lobbying in regard to atomic energy. Discusses the involvement of both organizations in the establishment of the National Science Foundation and the Atomic Energy Commission.

025. Klein, Randolph, editor. *Science and Society in Early America: Essays in Honor of Whitfield J. Bell, Jr.* Philadelphia: American Philosophical Society, 1986. 426 pp. Index.

Contains twelve articles. Includes items 392, 578, 585.

026. Kuznick, Peter J. *Beyond the Laboratory: Scientists as Political Activists in 1930s America.* Chicago: University of Chicago Press, 1987. x+363 pp. Index.

Views the 1930s as a period when American scientists appeared to resist the tendency to serve the existing power structure and accept the prevailing ideology. Documents increasing political activity against the status quo. Discusses the American interest in, and subsequent disillusionment with, the Soviet Union. Relates the history of the American Committee for Democracy and Intellectual Freedom and the American Association of Scientific Workers.

027. Mabee, Carlton. "Margaret Mead's Approach to Controversial Public Issues: Racial Boycotts in the AAAS." *The Historian,* 48 (1986): 191-208.

Looks at two occasions when the issue of boycotting the meeting of the American Association for the Advancement of Science was raised because of racial conflict. Demonstrates that Mead rejected a boycott and worked for change from within the system.

028. McMahon, A. Michal. "'Bright Science' and the Mechanic Arts: The Franklin Institute and Science in Industrial America, 1824-1976." *Pennsylvania History,* 47, (1980): 351-368.

Argues that changes in the Franklin Institute reflected larger changes in the place of technology and science in American culture.

029. McMurry, Linda O. *George Washington Carver: Scientist and Symbol.* Oxford: Oxford University Press, 1981. x+367 pp. Index.

> Assesses earlier scholarship. Separates fact from myth. Finds that Carver's fame rested more on his value as a symbol than any scientific accomplishments. Sees Carver being used as a symbol by many groups with contrary goals.

030. Mendelsohn, Everett, editor. *Transformation and Tradition in the Sciences: Essays in Honor of I. Bernard Cohen.* Cambridge: Cambridge University Press, 1984. xiv+577 pp. Index.

> Includes items 040, 107, 338.

031. Midgette, Nancy S. "The Role of the State Academies of Science in the Emergence of the Scientific Profession in the South, 1883-1983." Ph.D. dissertation, University of Georgia, 1984.

> Claims that academic scientists in the South organized state academies of science to increase peer contact because economic, transportation, and communication difficulties limited their active participation in the American Association for the Advancement of Science and the national scientific arena. Sees these difficulties as becoming less significant post-World War II, allowing southern scientists to take part in the national scientific community. Finds that the state academies of science have shifted their focus since World War II to the conservation of natural resources and the improvement of science education.

> Published as *To Foster the Spirit of Professionalism: Southern Scientists and State Academies of Science.* Tuscaloosa: University of Alabama Press, 1991. viii+238 pp. Bibliography, Index.

032. Numbers, Ronald L., and Janet S. Numbers. "Science in the Old South: A Reappraisal." *The Journal of Southern History,* 48 (1982): 163-184.

> Agrees with other research that science in the South lagged markedly behind the activity in the Northeast. Finds that southern activity increased during the decades just prior to the Civil War. Blames the lag on less urbanization and the limited availability of books, not the presence of slavery.

033. Pittenger, Mark. "Science, Culture and the New Socialist Intellectuals before World War I." *American Studies*, 28 (1987): 73-91.

Finds that this generation used scientific ideas as weapons in the conflict between differing factions among the socialists. Focuses on Walter Lippmann, Robert Lowie, and William E. Walling.

034. Pyne, Stephen J. *Fire in America: A Cultural History of Wildland and Rural Fire.* Princeton: Princeton University Press, 1982. xvi + 654 pp. Bibliography, Index.

Argues that humans have always modified their ecosystems through the use of fire. Traces the evolution of the concept of fire management by the United States Forest Service from fire prevention to the acceptance of burning under certain conditions. Intersperses chapters of regional history with those focusing on special themes, such as fire-fighting technology or fire research.

035. Reingold, Nathan, and Ida H. Reingold, editors. *Science in America: A Documentary History 1900-1939.* Chicago: University of Chicago Press, 1981. xii + 490 pp. Appendix, Index.

Organizes the documents by discipline, institution, or historical episode. Provides short introductory essays and light annotation. Conveys a sense of the professional life of a scientist.

036. Reingold, Nathan, and Joel N. Bodansky. "The Sciences, 1850-1900, a North Atlantic Perspective." *Biological Bulletin,* 168, Supplement (1985): 44-61.

Analyzes the official support of science by Germany, the United States, and the United Kingdom. Finds the overall patterns to be very similar. Credits the similarities to a tendency to emulate successful innovations in other countries. Concludes that most of the funds went for routine programs. Includes comparative tables.

037. Rice, Fredrica Dudley. "Emerson's Debt to Natural Science during His Early Life and Work." Ph.D. dissertation, University of Washington, 1984.

Finds five phases in Emerson's relationship to science, starting with his early and undergraduate education. Argues that Emerson transferred many of the assumptions of science into the realm of theology. Describes Emerson's life work as a "secular ministry based . . . on the progress of science."

038. Rosenberg, Charles. "Science in American Society: A Generation of Historical Debate." *Isis*, 74 (1983): 356-367.

Identifies three trends in the history of American science over the last twenty years: quantitative studies, the ideological uses of science, and the twentieth century and the rise of big science. Suggests that future historical research should look at the relationship of scientific ideas to the extra-scientific context through biographical studies of individuals or small groups.

039. Rossiter, Margaret W. *Women Scientists in America: Struggles and Strategies to 1940*. Baltimore: Johns Hopkins University Press, 1982. xviii+439 pp. Bibliography, Index.

Provides the starting point for future historical research on the role of gender in the scientific community. Contributes an inventory of the community of women scientists utilizing a database drawn from *American Men of Science*. Delineates two forms of discrimination which women scientists faced: hierarchical, which limited their advancement upward through what was alleged to be a meritocracy; and territorial, which limited them to certain fields. Discusses alternative reward systems developed by women.

040. Shigeru, Nakayama. "The American Occupation and the Science Council of Japan." *Transformation and Tradition in the Sciences: Essays in Honor of I. Bernard Cohen* (item 030), pp. 353-369.

Discusses the activities of Harry C. Kelly, who organized science in occupied Japan. Argues that Kelly wanted to establish a liaison with a representative group of Japanese scientists without including the prewar institutions. Sees Kelly's private advisory group--the Japan Association of Science Liaison--as the first step in a process of establishing increasingly more formal and public groups.

041. Stapleton, Darwin H. and Edward C. Carter, II. "'I have the itch of Botany, of Chemistry, of Mathematics . . . strong upon me': The Science of Benjamin Henry Latrobe." *Proceedings of the American Philosophical Society*, 128 (1984): 173-192.

Describes Latrobe's science as Baconian--descriptive and empirical. Presents him as a representative of the intellectual gentleman who formed the elite of American science during the Jeffersonian era. Characterizes his science as harmonious with the intellectual patterns of his age.

042. Thackray, Arnold. "On American Science." *Isis,* 73 (1982): 7-10.

Credits the increase in the quantity and quality of the work on the history of American science published since 1970 to the increasing interest of historians of science in more contemporary periods, periods in which American science was very important. Points also to practical considerations, such as language limitations, restricted travel budgets, and the interest of American scientific and technological institutions in their own history, in explaining the growing interest in American science.

043. Warner, Sam Ball, Jr. *Province of Reason.* Cambridge: Harvard University Press, 1984. ix + 302 pp. Index.

Recounts the lives of fourteen Bostonians (defined very broadly) active from 1870 to 1980. Focuses on the interplay between local circumstances, individual lives, and large national and international forces and events. Includes Vannevar Bush, Rachel Carson, James Bryant Conant, and Laurence K. Marshall. Expresses great unhappiness with the impact of modern science and technology on life.

044. Zernel, John Joseph. "John Wesley Powell: Science and Reform in a Positive Context." Ph.D. dissertation, Oregon State University, 1983.

Argues that Powell was convinced of the usefulness of science in achieving social progress. Shows that Powell believed that the failure of the government's policy towards Native Americans was a result of it not being based on a proper understanding of Native American culture.

GENERAL TECHNOLOGY

045. Bruins, Derk. "Technology and the Military: The Impact of Technological Change on Social Structure in the United States Navy." *Technology, the Economy, and Society: The American Experience* (item 047), pp. 223-250.

Warns that organizational change has deep social impact. Describes the resistance of the naval line officers to adapting the existing social structure to steam technology. Finds that technological change can result in changes in naval organization if the impact on the line officers is limited.

046. Clark, Jennifer. "The American Image of Technology from the Revolution to 1840." *American Quarterly*, 39 (1987): 431-449.

Argues that technology had a very positive image, and came to represent the dynamism of the new nation.

047. Colton, Joel, and Stuart Bruchey, editors. *Technology, the Economy, and Society: The American Experience.* New York: Columbia University Press, 1987. xiii+287 pp.

Looks at how segments of American society have attempted to cope with industrialization. Emphasizes the connections between the economy and technological change. Defines technology as the way energy is transformed and used. Includes items 045, 052, 074, 081, 108, 158, 500, 577, 627, 646.

048. Cooper, Carolyn, Robert Gordon, and Harvey Merrick. "Industrial Archeology at the Whitney Gun Factory Site." *Essays in Arts and Sciences,* 10 (1982): 135-149.

Summarizes the results of three archeological projects in the 1970s. Warns that excavations searching for building foundations will not provide evidence regarding Whitney's manufacturing methods. Argues that future researchers should look for trash dumps for such evidence.

049. Corn, Joseph J., editor. *Imagining Tomorrow: History, Technology, and the American Future.* Cambridge: MIT Press, 1986. vi+237 pp. Index.

Includes a series of case studies concerning the expectations Americans have had for the future impact of technology. Demonstrates that many predictions were wrong. Includes items 138, 508, 519, 525, 528, 539.

050. Donovan, Arthur L., editor. *Energy in American History. Materials and Society*, 7, No. 3&4 (1983): 243-487.

Contains twenty-two essays and commentaries on the history of fuels and energy policy. Includes contributions from historians of science and technology, business historians, and social scientists. Includes items 051, 084, 221, 550, 590, 619.

051. Giebelhaus, August W. "Petroleum's Age of Energy and the Thesis of American Abundance." *Energy in American History* (item 050), pp. 279-293.

Argues that petroleum has been abundant throughout most of the era in which it has been a major fuel source, but there have been numerous short-term shifts in availability. Asserts that the petroleum industry often was the first to exhibit characteristics which later were typical of much of American industry.

052. Hammack, David C. "Technology and the Transformation of the American Party System." *Technology, the Economy, and Society: The American Experience* (item 047), pp. 126-149.

Contends that changes in communication and transportation in the nineteenth century made possible the world's first stable party system based on a large electorate. Finds that twentieth-century technologies like the television and the airplane have assisted in the recent decline of the strength of the political parties because candidates can appeal over the heads of the party leadership directly to the voter.

053. Hart, Sidney. "'To Encrease the Comforts of Life': Charles Willson Peale and the Mechanical Arts." *Pennsylvania Magazine of History and Biography*, 110 (1986): 323-357.

Views Peale as an Enlightenment figure who thought that reason could be used to better the human condition. Discusses some of his inventions. Notes that all were commercial failures.

054. Hindle, Brooke. *Emulation and Invention.* New York: New York University Press, 1981. xx+162 pp. Index.

Combines pictorial and written essays in two case studies--the steamboat and the telegraph--exploring the mode of thinking which led to important technological innovations in the early nineteenth century. Focuses on the creative process, especially non-verbal forms of thinking. Speculates on the significance of spatial thinking in the evolution of machine design.

055. Hindle, Brooke. "'The Exhilaration of Early American Technology': A New Look." *The History of American Technology: Exhilaration or Discontent?* (item 059), pp. 7-17.

Reviews the literature in the history of American technology since 1965. Concludes that the field has grown and that networks of scholars have developed. Maintains that the literature of the history of science has been unimportant for the history of technology.

056. Hindle, Brooke, editor. *Material Culture of the Wooden Age.* Tarrytown: Sleepy Hollow Press, 1981. 394 pp. Bibliography, Index.

Contains ten papers, including items 509, 546, 589, 643. Focuses on the period through 1850. Looks at the use of wood in the home, farm, transportation, and production.

*   Hindle, Brooke. "A Retrospective View of Science, Technology, and Material Culture in Early American History." Cited above as item 023.

Reviews his work in history of technology. Sees the differences between traditional and industrial technology as a barrier that separates historians of technology. Note that there is a strong faith in the value of artifacts for the history of technology, but only occasional success.

057. Hounshell, David. "On the Discipline of the History of American Technology." *Journal of American History*, 67 (1981): 854-865.

Forecasts that the history of technology may either become part of social history or become focused on the internal dimensions and character of technology itself. Sees a danger that historians of technology, in an effort to denote disciplinary boundaries, may become more internal. Contrasts two different approaches to the history of technology as social history: starting from internal questions in the history of technology and moving out to issues of social history, versus starting with social history and viewing technology as part of the social experience. Item 078 is a response.

058. Hounshell, David. "The Discipline of the History of American
      Technology: An Exchange." *Journal of American History*,
      68 (1981-82): 900-902.

      Responds to item 078. Defends the idea of a discipline
      of the history of American technology within the larger context
      of American history.

059. Hounshell, David, editor. *The History of American Technology:
      Exhilaration or Discontent?* Wilmington: Hagley Museum
      and Library, 1984. 33 pp.

      Includes items 055, 072.

060. Jenkins, Reese V. "Words, Images, Artifacts and Sound:
      Documents for the History of Technology." *British
      Journal for the History of Science*, 20 (1987): 39-56.

      Argues that non-verbal documents are central to the
      historical and documentary editing enterprises in the history of
      technology. Offers a variety of examples from the papers of
      Thomas A. Edison. Emphasizes the use of photographs,
      drawings, and artifacts in understanding Edison's design process.

061. Kasson, John F. "The Invention of the Past: Technology,
      History, and Nostalgia." *Technological Change and the
      Transformation of America*. Edited by Steven E. Goldberg
      and Charles R. Strain. Carbondale: Southern Illinois
      University Press, 1987, pp. 37-52.

      Argues that nostalgia has replaced a sense of history in the
      United States. Blames this replacement on America's develop-
      ment as a technological society and the rise of the consumer
      culture. Warns that Americans view history as a progressive
      development of consumer goods and inventions, not as the record
      of human experiences. Views Disney's Main Street as an
      example of history devoid of substance--of an idealized past.

062. Kranakis, Eda Fowlks. "Technological Styles in America and
      France in the Early Nineteenth Century: The Case of the
      Suspension Bridge." Ph.D. dissertation, University of
      Minnesota, 1982.

      Views technological styles--which include both traditions
      of design and technological knowledge--as an interface between
      the internal and external aspects of the history of technology.
      Uses James Finley as a case study. Contrasts the American style,

which stressed functionalism in design and an empirico-inductive research methodology, with the French state engineers, who used a theoretico-deductive methodology. Finds that the style of French private sector engineers more closely resembles that of Americans than their countrymen in the public sector.

Partially published as "Social Determinants of Engineering Practice: A Comparative View of France and America in the Nineteenth Century." *Social Studies of Science*, 19 (1989): 5-70.

063. Lubar, Steven David. "Corporate and Urban Contexts of Textile Technology in Nineteenth-Century Lowell, Massachusetts: A Study of the Social Nature of Technological Knowledge." Ph.D. dissertation, University of Chicago, 1983.

Studies both the large corporations of Lowell and innovation and entrepreneurship in the larger community. Defines technology as a type of social knowledge. Identifies three forms of technological knowledge: knowledge that people have, knowledge contained in machines and structures, and knowledge contained in organizations.

064. McGaw, Judith A. "Accounting for Innovation: Technological Change and Business Practice in the Berkshire County Paper Industry." *Technology and Culture*, 26 (1985): 703-725.

Advocates the recognition of the history of accounting as an aspect of the history of technology. Defends the accounting systems used by the paper mill owners as appropriate for their needs. Finds that accounting procedures changed as needs demanded.

065. Marx, Leo. "Does Improved Technology Mean Progress?" *Technological Change and the Transformation of America.* Edited by Steven E. Goldberg and Charles R. Strain. Carbondale: Southern Illinois University Press, 1987, pp. 23-36.

Argues that the criterion of progress has undergone a major change during the last two centuries. Believes that the Founding Fathers saw a linkage between increasing control over nature and increasing political freedom. However, by the 1840s, scientific and technical progress had been disassociated from the political vision of progress. Innovations in science and technology were in themselves indications of progress, irrespective of the societal ends. Views the Transcendental writers as critics of the new view of progress.

066. Melosi, Martin V. *Coping with Abundance: Energy and Environment in Industrial America, 1820-1980.* Philadelphia: Temple University Press, 1985. xii+355 pp. Index.

Sees abundant energy as the distinguishing characteristic of the industrial United States. Identifies two major transitions: from renewable energy sources--wood, water, and wind--to non-renewable--coal--in the nineteenth century, and from coal to oil in the twentieth. Shows that both transitions were gradual. Emphasizes the period from World War I to the 1970s.

067. Musgrove, Susan M. "Charles A. Lindbergh and the American Dilemma: Tension and Resolution in the Conflict between Technology and Human Values." Ph.D. dissertation, St. Louis University, 1985.

Uses the life of Lindbergh to illuminate the three stages in the relationship between technology and human values in twentieth–century America: technology seen as a boon to humanity; technology viewed in terms of its destructive force; and finally, the two seen as complementary.

Published as Susan M. Gray, *Charles A. Lindbergh and the American Dilemma: The Conflict of Technology and Human Values.* Bowling Green, Ohio: Bowling Green State University Popular Press, 1988. 128 pp. Bibliography, Index.

068. Nye, David E. *Image Worlds: Corporate Identities at General Electric, 1890-1930.* Cambridge: MIT Press, 1985. xiv+188 pp. Index.

Looks at the corporate photography collection at General Electric as a means of studying the corporation as a communicator during the period after the rise of commercial photography and before commercial radio. Finds that the photograph presented ideologies selected by management for one of four audiences: engineers, workers, managers, and consumers.

069. Nye, David E. *The Invented Self: An Anti-Biography, from Documents of Thomas A. Edison.* Odense: Odense University Press, 1983. 229 pp. Index.

Rejects biography and narrative. Utilizes the ideas of structuralists and semioticians. Rejects a chronological approach. Focuses on specific points in Edison's life. Demonstrates, unintentionally, the limitations of semiotic approaches to historical research.

070. Post, Robert, editor. *American Enterprise: Nineteenth-Century Patent Models*. New York: Cooper-Hewitt Museum, 1984. 143 pp. Bibliography, Chronology.

Sees patent models as providing an artifactual record of the great growth of technology innovation in the United States during the years 1836-1880. This is a heavily illustrated exhibition catalogue.

071. Pursell, Carroll W., Jr. "The History of Technology and the Study of Material Culture." *American Quarterly*, 35 (1983): 304-315.

Surveys the connections and importance of studying technology through objects. Argues that the responsibility of the history of technology is to decipher the culture embedded in material objects.

072. Pursell, Carroll W., Jr. "The Problematic Nature of Late American Technology." *The History of American Technology: Exhilaration or Discontent?* (item 059), pp. 18-27.

Reviews the literature since 1965 for the history of American technology after 1850. Concludes that the nature of modern technology has resulted in less exhilaration among scholars of this era than their counterparts for the earlier period in American history.

073. Pursell, Carroll W., Jr., editor. *Technology in America: A History of Individuals and Ideas*. Cambridge: The MIT Press, 1981. xi+264 pp. Bibliography, Index.

Contains twenty brief, undocumented essays; some are précis of earlier publications, others are new synthetic works. Focuses on individuals. Includes items 077, 087, 489, 499, 537, 548, 557.

074. Scheiber, Harry N. "The Impact of Technology on American Legal Development, 1790-1985." *Technology, the Economy, and Society: The American Experience* (item 047), pp. 83-125.

Argues that changes in technology forced a response on the part of the legal system, and in turn, the law influenced the diffusion of technology. Sees the legal system as very sympathetic to technology in the nineteenth century, spreading the costs of innovation throughout the community. Finds that the courts have justified the expansion of the power of government in the twentieth century by arguing that government had to be able to respond to changing technology. Believes that the legal system encourages litigation over patent claims.

075. Sinclair, Bruce, editor. *New Perspectives on Technology and American Culture.* Philadelphia: American Philosophical Society, 1986. ix + 80 pp.

Contains five papers discussing technology and technologists from the perspective of American cultural history. Includes items 092, 507, 564, 565.

076. Stabile, Donald. *Prophets of Order: The Rise of the New Class, Technocracy and Socialism in America.* Boston: South End Press, 1984. 295 pp. Index.

Focuses on the period 1890-1930. Defines the "New Class" as the experts who are able to understand the machines upon which technological American depends. Places these individuals between the workers and the capitalists. Sees them as seeking a niche within capitalism by claiming power through expertise. Argues that some members of the New Class wanted to be scientific mediators of class conflict. Distinguishes between the "technical intelligentsia," chiefly engineers, who sought to rationalize the work place, and "social intellectuals, who wanted to give direction to the country as a whole.

077. Stapleton, Darwin H. "Benjamin Henry Latrobe and the Transfer of Technology." *Technology in America: A History of Individuals and Ideas* (item 073), pp. 34-44.

Shows how a single individual transferred European technology to the United States through his own knowledge, his training of pupils, his employment of immigrants, and his participation in a community of immigrants and Americans who had visited Europe.

078. Stapleton, Darwin H. "The Discipline of the History of American Technology: An Exchange." *Journal of American History,* 68 (1981-82): 897-900.

Responds to item 057. Rejects the view that the history of technology should have a closer relationship with social history. Expresses the belief that few scholars in the history of technology define their interests in terms of national boundaries.

079. Stapleton, Darwin H. *The Transfer of Early Industrial Technologies to America. Memoirs of the American Philosophical Society,* 177. Philadelphia: American Philosophical Society, 1987. x +215 pp. Bibliography, Index.

Summarizes recent research. Provides four case studies from the Mid-Atlantic region: William Weston, Benjamin Latrobe, and Philadelphia internal improvements; E. I. du Pont and black powder; Moncure Robinson and railroad technology; and David Thomas and anthracite iron industry. Finds that support for technological transfer was provided by the government, the financial community, and the mercantile community. Concludes that the United States was receptive to technological innovation.

080. Tarr, Joel A., et al. "Water and Wastes: A Retrospective Assessment of Wastewater Technology in the United States, 1800-1932." *Technology and Culture,* 25 (1984): 226-263.

Focuses on the urban environment. Concludes that the exacting secondary costs of apparently beneficial innovations were difficult to anticipate. Discovers that faulty scientific concepts led to faulty technological decisions.

081. Vatter, Harold G. "Technological Innovation and Social Change in the United States, 1870-1980." *Technology, the Economy, and Society: The American Experience* (item 047), pp. 19-55.

Sees economic change as the link between technological innovation and social change. Distinguishes fourteen aspects of American society which were effected by increased per capita income resulting from technological innovation in manufacturing, power, and transportation. Designates World War I as the important transition point.

082. Wachhorst, Wyn. *Thomas Alva Edison: An American Myth.*
      Cambridge: M.I.T. Press, 1981. x+328 pp. Appendix,
      Bibliography, Index.

      Chronicles the development of a cultural myth. Argues
      that at the end of the nineteenth-century Edison's image took on
      a life separate from the man. Sees Edison as a component of the
      American myth of dependency upon the accomplishments of the
      lone hero.

083. Weber, Daniel B. "'The Manufacturer and Builder': Science,
      Technology, and the American Mechanic." *Journal of
      American Culture,* 8, No. 4, Winter 1985: 35-42.

      Characterizes the journal as a nineteenth-century version
      of *Popular Science* or *Popular Mechanics.* Describes it as a post-
      Civil War affirmation that technology was the American way of
      life, that industrialism leads to national perfection, and science
      and technology were vehicles for human progress.

084. Wilbanks, Thomas J. "Geography and Our Energy Heritage."
      Energy in American History (item 050), pp. 437-452.

      Presents the perspective of a geographer. Identifies three
      topics where energy intersects with research traditions in
      geography: the character of regions of the United States, energy
      and spatial structure, and energy and landscapes. Contends that
      energy use has been more important to American history than
      energy supply.

085. York, Neil L. *Mechanical Metamorphosis: Technological Change
      in Revolutionary America.* Westport: Greenwood Press,
      1985. xviii+240 pp. Bibliography, Index.

      Traces the changing American attitude towards technology
      during the Revolutionary War period. Provides selective case
      studies. This is a revised version of item I:148.

CHAPTER II: SPECIAL THEMES

ART AND LITERATURE

086. Ammons, Elizabeth. "The Engineer as Cultural Hero and Willa Cather's First Novel, *Alexander's Bridge.*" *American Quarterly,* 38 (1986): 746-760.

Sees Cather's novel as attacking the myth of the engineer as the educated man of action, able to combine adventure and accomplishment as he brought civilization to the wilderness. Describes Cather's hero as the epitome of the engineer image at the turn of the century.

087. Basalla, George. "Keaton and Chaplin: The Silent Film's Response to Technology." *Technology in America: A History of Individuals and Ideas* (item 073), pp. 192-201.

Summarizes Keaton's attitude towards machines as wary; the needs of machines and humans do not intertwine smoothly. Sees the theme of Chaplin's *Modern Times* as the inability of humans to master machines.

088. Blum, Ann Shelby. "'A Better Style of Art': The Illustrations of the *Paleontology of New York.*" *Earth Sciences History,* 6 (1987): 72-85.

Views scientific illustration as representing both translation and abstraction. Presents a case study of the process of rendering a specimen into first a drawing and then a printed plate. Traces the evolution of the production of scientific illustration from a family affair to a profession.

089. Cooper, Peter L. *Signs and Symptoms: Thomas Pynchon and the Contemporary World.* Berkeley: University of California Press, 1983. x+238 pp. Bibliography, Index.

Studies Pynchon in terms of his literary, philosophical, and scientific contexts. Sees his fiction as a reformulation, in human terms, of issues raised in modern theoretical physics, including entropy, quantum theory, and Heisenberg's uncertainty principle.

090. Handler, Richard. "Vigorous Male and Aspiring Female: Poetry, Personality, and Culture in Edward Sapir and Ruth Benedict." *Malinowski, Rivers, Benedict and Others: Essays on Culture and Personality* (item 439), pp. 127-155.

Uses the aesthetic philosophy of poets, especially Ezra Pound's metaphor of hardness, to understand the differing views of Sapir and Benedict toward the relationship of the individual to culture. Defines hardness as a male culture attribute: "self standing alone." Contrasts Sapir's need to accommodate the role of family with Benedict's need to decide between masculine and feminine aspirations.

091. Harris, Neil. "The Drama of Consumer Desire." *Yankee Enterprise: The Rise of the American System of Manufactures* (item 609), pp. 189-216.

Looks at the growth of mass-produced objects and consumption as symbol and metaphor in American novels through the interwar years.

092. Horrigan, Brian. "Popular Culture and Visions of the Future in Space, 1901-2001." *New Perspectives on Technology and American Culture* (item 075), pp. 49-67.

Surveys the theme of space travel in popular media. Finds that there is a general agreement on the role of technology in the conquest of space. Discusses the mixed and changing rationales for space travel, including resource exploitation, the thrill of exploration, and political or military domination.

093. Kranzberg, Melvin. "Confrontation or Complementarity?: Perspectives on Technology and the Arts." *Bridge to the Future: A Centennial Celebration of the Brooklyn Bridge* (item 558), pp. 333-345.

Argues that art and technology are different but complementary facets of human culture. Describes technology's contribution to art in terms of tools and inspiration for artists; art is part of the broad social context which influences technology. Notes the aesthetic element of technology.

094. Krieger, William Carl. "Henry David Thoreau and the Limitations of Nineteenth-Century Science." Ph.D. dissertation, Washington State University, 1986.

Studies the early works of Thoreau, focusing on the period prior to that discussed in item 097. Believes Thoreau rejected Baconian science because it excluded insight and philosophy. Describes Thoreau as a follower of an alchemical, elemental approach.

095. Limon, John Keith. "Imagining Science: The Influence and Metamorphosis of Science in Charles Brockden Brown, Edgar Allan Poe, and Nathaniel Hawthorne." Ph.D. dissertation, University of California, Berkeley, 1981.

Shows how these writers used current scientific world views in their writing. Argues that during the sixty years covered by the combined careers of these writers, there was a falling-off in the ability of writers to understand the science.

Published in part in *The Place of Fiction in the Time of Science: A Disciplinary History of American Writing*. Cambridge: Cambridge University Press, 1990. xiii+216 pp. Index.

096. Martin, Ronald E. *American Literature and the Universe of Force*. Durham: Duke University Press, 1981. xviii+284 pp. Bibliography, Index.

Studies the concept of force-universe, as derived from the thought of Herbert Spencer, as it appeared in science, philosophy, and literature, from its origination, in the mid-nineteenth century, until its assimilation into American literature, approximately 1900. Provides analysis of the writings of Henry Adams, Frank Norris, Jack London, and Theodore Dreiser.

097. Rossi, William John. "'Laboratory of the Artist': Henry Thoreau's Literary and Scientific Use of the Journal, 1848-1854." Ph.D. dissertation, University of Minnesota, 1986.

Sees Thoreau's journal composition as serving two purposes: producing a draft for literary works and a record of and reflection on his nature observations. Presents Thoreau as a critic of positivist methodology in science. For a study of Thoreau's earlier work, see item 094.

098. Segal, Howard P. *Technological Utopianism in American Culture.* Chicago: University of Chicago Press, 1985. x + 301 pp. Appendix, Bibliography, Index.

Studies twenty-five technological utopias delineated in literature published between 1883 and 1933. Describes the vision projected in this literature as an acceleration of contemporary trends. Includes a very extensive and useful bibliography.

099. Simons, Kent Steven. "Taming the Powers of the Air: Science, Pseudoscience, and Religion in Nineteenth-Century American Literature." Ph.D. dissertation, Emory University, 1985.

Looks at the great range of responses of American writers, especially Poe and Twain, to discoveries in nineteenth-century science. Focuses on their treatment of the relationship between mind and spirit. Argues that several of their themes are suggestive of traditional religious thought.

100. Slade, Joseph W. "Thomas Pynchon, Postindustrial Humanist." *Technology and Culture*, 23 (1982): 53-72.

Claims that Pynchon is the first American novelist to express the nature of modern technology and the crisis it has spawned.

101. Sloane, David E. E. "Connecticut Yankee and Industrial America: Mark Twain's Lesson." *Essays in Arts and Sciences*, 10 (1982): 197-205.

Interprets Twain as envisioning industrialism and technology in humanistic and patriotic terms.

102. Steinman, Lisa M. *Made in America: Science, Technology, and American Modernist Poets.* New Haven: Yale University Press, 1987. xiv + 219 pp. Index.

Focuses on modernist poetry and its defenses, especially the work of William Carlos Williams, Marianne Moore, and

Wallace Stevens between 1910 and 1945. Argues that science and technology were often used interchangeably or lumped together until at least 1920. Emphasizes the role of modern physics, especially Albert Einstein's theories, in the thought of these poets. Sees the poets as developing a poetic drawing analogies to science in order to defend their style and justify their position in American society. Emphasizes that the poets became aware of scientific theories through popularizers rather than direct contact with the scientific literature.

103. Stilgoe, John R. "Fair Fields and Blasted Rocks: American Land Classification Systems and Landscape Aesthetics." *American Studies,* 22, No.1 (Spring 1981): 21-33.

 Argues that agricultural and mineralogical land-classification systems influenced the way American painters and writers treated landscapes.

104. Tichi, Cecelia. *Shifting Gears: Technology, Literature, Culture in Modernist America.* Chapel Hill: University of North Carolina Press, 1987. xvii+310 pp. Bibliography, Index.

 Sees the technological revolution of the period 1890-1920 as also one of language, fiction and poetry. Argues that the gear and girder not only became dominant as metaphors in literature, but that fiction took the form of a mechanical system designed to deliver force. Concentrates on elite literature. Contends that William Carlos Williams defined a machine-age poetics with his rapid cadence. Presents a very controversial analysis.

105. Wade, Edwin L. "The Ethnic Art Market in the American Southwest, 1880-1980." *Objects and Others: Essays on Museums and Material Culture* (item 440), pp. 167-191.

 Divides the century into three periods: during 1875-1915 economic bonds were created between the Indian traders and the scholars, who depended upon the former for artifacts; the Art Revivalist Movement, 1920-1970, was marked by art patrons driving a wedge between dealers and scholars, and the efforts by philanthropists to revive native artistic traditions; and post-1968, the reorientation of the market.

EDUCATION

106. Arnold, Lois Barber. *Four Lives in Science: Women's Education in the Nineteenth Century.* New York: Schocken, 1984. xii + 179 pp. Index.

    Offers biographies of Maria Martin Bachman, Almira Hart Lincoln Phelps, Louisa C. Allen Gregory, and J. Florence Bascom, representing examples of women receiving increasingly more formal scientific education as the century progressed. Presents the education of these women as typical of that of American female scientists of the nineteenth century. Equates her study with the findings of Clark Elliott (Item I:39).

107. Buck, Peter S., and Barbara Gutmann Rosenkrantz. "The Worm in the Core: Science and General Education." *Tradition in the Sciences: Essays in Honor of I. Bernard Cohen* (item 030), pp. 371-394.

    Assesses the failure, during the twentieth century, to include science as an indispensable element of the general college curriculum. Argues that the failure was due to the inability to agree on the relevance of scientific knowledge to human affairs.

108. Clifford, Geraldine Joncich. "The Impact of Technology on American Education, 1880-1980." *Technology, the Economy, and Society: The American Experience* (item 047), pp. 251-277.

    Concludes that technological innovations have had many indirect effects on education, but that instructional technology has had little impact on the activity in the classroom.

109. Owens, Larry. "Pure and Sound Government: Laboratories, Playing Fields, and Gymnasia in the Nineteenth-Century Search for Order." *Isis*, 76 (1985): 182-194.

    Explores the changing culture of American colleges through the perspectives of the university as a laboratory (Johns Hopkins), playing fields as an alternative republic within academia (Yale), and, through the gymnasia, the school as an ordered, well-balanced organism (Harvard). Perceives laboratories, playing fields, and gymnasia as alternative means of imposing control and providing a forum for a new form of order in the post-bellum college.

ENVIRONMENT AND CONSERVATION

110. Burdick, Neal Stephens. "The Evolution of Environmental
     Consciousness in Nineteenth-Century America: An
     Interdisciplinary Study." Ph.D. dissertation, Case Western
     Reserve University, 1981.

     Traces the change in attitude towards the environment--
     from negative to positive--in the writings and work of James
     Fenimore Cooper, Henry David Thoreau, George Perkins Marsh,
     Verplanck Colvin, and John Muir. Finds in this work the
     intellectual roots of both twentieth-century conservation ideology
     and environmental activism.

111. Cohen, Michael P. *The Pathless Way: John Muir and American
     Wilderness*. Madison: University of Wisconsin Press,
     1984. xvii+408 pp. Index.

     Views Muir as more radical in his thinking regarding the
     relationship between humans and nature than traditionally inter-
     preted from his published writings. Utilizes Muir as the focal
     point in a broader analysis of the human response to nature and
     the wilderness. Interjects preservationist attitudes into the
     historical discussion. Attacks the history of conservation as
     anthropocentric.

112. Cutright, Paul Russell. *Theodore Roosevelt: The Making of a
     Conservationist*. Urbana: University of Illinois Press,
     1985. xiii+287 pp. Bibliography, Index.

     Provides a biography of Roosevelt which emphasizes his
     naturalist activities. Argues that Roosevelt's sympathy for
     conservation was due to his youthful natural history interests and
     studies.

113. Dunlap, Thomas R. *DDT: Scientists, Citizens, and Public Policy*.
     Princeton: Princeton University Press, 1981. 318 pp.
     Appendices, Bibliography, Index.

     Traces the history of DDT from the discovery of its
     insecticidal properties, through the debates over its safety, the
     role of these debates in the establishment of the Environmental
     Defense Fund, and the banning of its use on agricultural crops
     in 1972 by the Environmental Protection Agency. Observes that
     the research programs of economic entomology changed
     enormously in the wake of the battle over DDT. Credits the

battle with giving birth to the integrated pest management approach.

114. Egerton, Frank N. "The History of Ecology: Achievements and Opportunities, Part Two." *Journal of the History of Biology,* 18 (1985): 103-143.

> Surveys the literature in the history of applied ecology in the United States and Canada. Distinguishes between applied ecology and conservation. Divides the discussion into four major areas: agriculture and forestry, fisheries management, wildlife management, and public health and environmental protection and planning. Lists approximately three hundred publications.

115. Fox, Stephen. *John Muir and His Legacy: The American Conservation Movement.* Boston: Little, Brown, 1981. xii+436 pp. Bibliography, Index.

> Combines a biography of Muir with a history of the conservation movement emphasizing the role of amateur radicals like Muir. Provides new understanding of Muir. Emphasizes the polarities of his life.

116. Hays, Samuel P. *Beauty, Health, and Permanence: Environmental Politics in the United States, 1955-1985.* Cambridge: Cambridge University Press, 1987. xv+630 pp. Index.

> Provides a topical rather than chronological discussion. Perceives the postwar environmental movement as very different from the prewar. Describes the environmental movement as a consumer movement, concerned about products which have aesthetic and restorative uses, arising out of an increase in the standard of living and the desire of the middle class for better living conditions. Believes that it is a local, innovative, grassroots movement. Criticizes, sometimes unfairly, the actions of public resource managers and scientists, economists, and other holders of expert knowledge.

117. Kelley, Donald Brooks. "Friends and Nature in America: Toward an Eighteenth-Century Quaker Ecology." *Pennsylvania History,* 53 (1986): 257-272.

> Describes the Quaker belief in man as custodian of God's environment, which led to a "moral ecology."

118. Levine, Richard. "Indians, Conservation, and George Bird Grinnell." *American Studies,* 28, No. 2 (Fall 1987): 41-55.

Surveys Native American attitudes toward nature and Euroamerican interpretations of those attitudes. Uses Grinnell as a source for conservation attitudes toward Native Americans around 1900.

119. Mighetto, Lisa. "Wild Animals in American Thought and Culture, 1870s-1930s." Ph.D. dissertation, University of Washington, 1986.

Argues that during this period anthrocentrism--the philosophy that the world was created for the use of humans--was displaced by a concern for preserving the quality of life of wild animals. Identifies a number of motivations for this new concern, including the "Back to Nature" movement as a response to the rapid industrialization of America, worry about individual animals, and the belief that animals had rights.

Published as *Wild Animals and American Environmental Ethics.* Tucson: University of Arizona Press, 1991. xiv + 177 pp.

120. Pisani, Donald J. "Forests and Conservation, 1865-1890." *The Journal of American History,* 72 (1985): 340-359.

Argues that a conservation ethic existed before the rise of Progressive reform. Finds the ethic grew out of a fear of a shortage of timber.

121. Smith, Michael L. *Pacific Visions: California Scientists and the Environment, 1850-1915.* New Haven: Yale University Press, 1987. ix + 243 pp. Index.

Focuses on a few key earth or life scientists active in the San Francisco Bay area. Identifies the goal of these scientists as the nurturing of an environmentally literate public. Contends that California scientists emphasized practices and values either absent or being actively rejected by Atlantic seaboard scientists as part of professionalization. Among these were an environmental awareness, a continuing interest in observational science, and a concern for a public role for science. Concludes that by World War I, California scientists had been integrated into the national community.

122. Stone, Carol Beall Leth. "From Forests to Fields to Food Webs: The Environment in History and Biology Textbooks, 1905-1975." Ph.D. dissertation, Stanford University, 1984.

Follows the changing portrayals of conservation and ecology in biology and history textbooks. Identifies among the changes in biology textbooks a shift in terminology from "conservation" to ecology," an increasing level of abstraction in the discussions, and a shift after 1960 to a more ecocentric perspective; in contrast, history textbooks have remained anthropocentric. Finds that in both types of textbooks, values have become more implicit and responsibility for behavior has shifted from the individual to the group, and ultimately, to government and experts.

123. Stott, R. Jeffrey. "The American Idea of a Zoological Park: An Intellectual History." Ph.D. dissertation, University of California, Santa Barbara, 1981.

Contends that the zoo at the turn of the twentieth century was shaped by a number of intellectual currents: conservation, City-Beautiful, scientific efficiency, imperialism, and concerns about immigration.

EXPEDITIONS

124. Engstrand, Iris H. W. *Spanish Scientists in the New World: The Eighteenth-Century Expeditions.* Seattle: University of Washington Press, 1981. xiv+220 pp. Appendices, Bibliography, Index.

Focuses on the Royal Scientific Expedition to New Spain and the expedition of Alejandra Malespina. Finds that the Spanish were most interested in practical returns from their scientists.

125. Goetzmann, William H., and Kay Sloan. *Looking Far North: The Harriman Expedition to Alaska, 1899.* New York: Viking, 1982. xxv+244 pp. Appendix, Bibliography, Index.

Provides a detailed chronological account of the last of the grand exploring cruises of the Second Great Age of Discovery. Attacks the expedition members for their indifference to the exploitation of the Alaskan wilderness. Evaluates the scientific contributions of the expedition in the last chapter.

126. Viola, Herman J., and Carolyn Margolis, editors. *Magnificent Voyagers: The U. S. Exploring Expedition, 1838-1842.* Washington, D.C.: Smithsonian Institution Press, 1985. 303 pp. Index, Appendices.

Contains twelve articles, five on the scientific work, five on the naval and diplomatic aspects of the expedition, and two on the fate of the collections in Washington, D.C.

PATRONAGE

127. Abir-Am, Pnina. "The Discourse of Physical Power and Biological Knowledge in the 1930s: A Reappraisal of the Rockefeller Foundation's 'Policy' in Molecular Biology." *Social Studies of Science,* 12 (1982): 341-382.

Distinguishes between the concepts of patronage and policy. Relies heavily on Michel Foucault's definition of power in developing its thesis. Looks at the impact of Rockefeller Foundation policy on three scientific projects important in the rise of molecular biology--William T. Astbury's work on x-ray studies of biological tissue, Linus Pauling's search for the "Secret of Life," and the work of the Institute of Mathematico-Physico-Chemical Morphology. Contends that the Rockefeller Foundation's policy favored established individuals at prestigious institutions and institutionally secure projects, and utilized a definition of eligibility which allowed physical scientists to enter biology without ever developing an interest in biology's conceptual problems. This is a highly controversial article which drew a number of responses, which were published, along with the author's reply, in *Social Studies of Science,* 14 (1984): 225-263.

128. Baatz, Simon. "Patronage, Science, and Ideology in an American City: Patrician Philadelphia, 1800-1860." Ph.D. dissertation, University of Pennsylvania, 1986.

Argues that the city's patriciate supported science in an effort to preserve the city's status in national culture. Correlates the ability of new scientific organizations to obtain local patronage with cooperative attitudes towards the patriciate and their scientific organizations. Finds that the local scientific leadership gave little support to the national American Association for the Advancement of Science.

Partially published in "Philadelphia Patronage: The Institutional Structure of Natural History in the New Republic, 1800-1833." *Journal of the Early Republic,* 8 (1988): 111-138.

129. Geiger, Roger L. *To Advance Knowledge: The Growth of American Research Universities, 1900-1940.* New York: Oxford University Press, 1986. x+325 pp. Appendices, Bibliography, Index.

Focuses on sixteen elite universities which had the majority of the leading scientists in the United States during the period. Sees their ability to raise funds as a common denominator. Interweaves the history of the universities with those of their major funding sources--foundations, business, and the government.

130. James, Mary Ann. *Elites in Conflict: The Antebellum Clash over the Dudley Observatory.* New Brunswick: Rutgers University Press, 1987. xiii+301 pp. Bibliography, Indices.

Views the clash between the lay trustees and the Scientific Council of the Dudley Observatory as one between two elites, each seeking deference, each responding to its own sense of responsibility. Rejects the interpretation set forth in Item I:406.

131. Kohler, Robert E. "Science, Foundations, and American Universities in the 1920s." *Osiris,* 2nd ser., 3 (1987): 135-164.

Identifies the partnership between private philanthropy and the university as one of the most distinctive features of American science between the World Wars. Argues that the catalyst was the linkage of research to the expansion of graduate training. Focuses on the Carnegie and Rockefeller programs.

132. Lankford, John. "Private Patronage and the Growth of Knowledge: The J. Lawrence Smith Fund of the National Academy of Sciences, 1884-1940." *Minerva,* 25 (1987) 269-281.

Illustrates the process of private support for research at a time of small-scale scientific research through the study of a fund for the support of meteoritic science. Concludes that the availability of funding is not sufficient to hasten the ripening process of a field of research.

133. Stanfield, John H. *Philanthropy and Jim Crow in American Social Science*. Westport, Conn.: Greenwood Press, 1985. xii+216 pp. Bibliography, Index.

Argues that the sponsorship by major foundations of sociological research on African-Americans derived more from issues of sociopolitical control than from a desire for empirical truth. Accuses the foundations of creating a knowledge base which accommodated Jim Crow laws and traditions. Maintains that the philanthropists and the foundation administrators saw knowledge as a means of social control, and sought to implement their ideas by supporting specific scientists. Focuses on the Laura Spelman Rockefeller Memorial, the Julius Rosenwald Fund, and the Carnegie Corporation.

134. Stocking, George W., Jr. "Philanthropoids and Vanishing Cultures: Rockefeller Funding and the End of the Museum Era in Anglo-American Anthropology." *Objects and Others: Essays on Museums and Material Culture* (item 440), pp. 112-145.

Argues that the Rockefeller Foundation played a major role in the reorientation of anthropology between the World Wars by providing funds to university-based researchers rather than museum staff and supporting the social science orientation of anthropology. Rejects the argument that ideological agendas shaped the Rockefeller program. Focuses on the contributions of Beardsley Ruml, Edwin R. Embree, and Edmund Day.

135. Turner, Stephen P. "The Survey in Nineteenth-Century American Geology: The Evolution of a Form of Patronage." *Minerva*, 25 (1987): 282-330.

Sees the geological survey as a major step in the evolution of the scientific patronage system in the United States from support for scientists as clients, in the tradition of the spoils system, to widespread support of science, legitimated by the authority of the scientific community and the benefits of scientific research. Credits John Wesley Powell with persuading political leaders, acting as patrons, that scientists should be judged by their peers rather than by politicians utilizing political or personal standards.

POPULAR SCIENCE

136. Burnham, John C. *How Superstition Won and Science Lost: Popularizing Science and Health in the United States.* New Brunswick: Rutgers University Press, 1987. xi + 369 pp. Index.

    Defines popularization as the translation of scientific discoveries into nonscientific popular language. Focuses on the late nineteenth and the twentieth centuries. Finds that nineteenth and early twentieth-century popularizers also tried to persuade non-scientists of the validity of scientists' methodology and worldview. Discusses the withdrawal of scientists from popularization efforts, leaving the activity to media specialists. Describes the efforts of the modern popularizes as the presentation of facts in isolation and appeals to authority.

137. Caudill, Charles Edward. "The Evolution of an Idea: Darwin in the American Press, 1860-1925." Ph.D. dissertation, University of North Carolina at Chapel Hill, 1986.

    Examines the reaction in selected newspapers and magazines to Darwin's theory of evolution. Sees the press as focusing on conflicts. Concludes that the press frequently failed to transmit complex ideas very well.

    Published as *Darwinism in the Press: The Evolution of an Idea.* Hillsdale, N.J.: Erlbaum, 1989. xvi + 161 pp. Bibliography, Index.

138. Kihlstedt, Folke T. "Utopia Realized: The World's Fairs of the 1930s." *Imagining Tomorrow: History, Technology and the American Future* (item 049), pp. 97-118.

    Argues that the Chicago and New York World's Fairs suggested that corporate capitalism would lead to utopia.

139. Kuritz, Hyman. "The Popularization of Science in Nineteenth-Century America." *History of Education Quarterly*, 21 (1981): 259-274.

    Links the popularization of science in the United States with the democratic movement. Connects popularization to the concept of self-improvement for the skilled worker.

140. Lewenstein, Bruce V. "'Public Understanding of Science' in America, 1945-1965." Ph.D. dissertation, University of Pennsylvania, 1987.

    Looks at the rising formal, institutional concern with popular science by commercial publishers, scientific societies, science writers, and government agencies. Concludes that a consensus arose equating public understanding of science with "public appreciation of and support for the benefits that science provides society." Credits that consensus to shared values among the four groups, including a dedication to the value of science in the modern world and a conviction that science and democracy were complementary.

    Partially published in "Magazine Publishing and Popular Science after World War II." *American Journalism,* 6 (1989): 218-234.

141. Post, Robert C. "Reflections of American Science and Technology at the New York Crystal Palace Exhibition of 1853." *Journal of American Studies,* 17 (1983): 337-356.

    Contends that the displays at the Crystal Palace reflected contemporary American technology, emphasizing transportation, communication, measurement, the mechanization of production, and the protection of persons and property. Finds little difference in the self-promotion techniques utilized by scientists and inventor-entrepreneurs.

142. Reingold, Nathan. "Metro-Goldwyn-Mayer Meets the Atom Bomb." *Expository Science: Forms and Functions of Popularisation.* Edited by Terry Shinn and Richard Whitley. *Sociology of the Sciences,* 9 (1985): 229-245.

    Discusses the evolution of the script of *The Beginning or the End,* a film released in 1947 on the building of the atomic bomb. Suggests that the result of the episode may have been an increasing concern by scientists to reach the lay public.

143. Rhees, David Jerome. "The Chemists' Crusade: The Rise of an Industrial Science in Modern America, 1907-1922." Ph.D. dissertation, University of Pennsylvania, 1987.

    Traces the function of popularization in obtaining industrial support for chemistry during the period from the publication of Robert Kennedy Duncan's *The Chemistry of Commerce* to the passage of the Fordney-McCumber tariff act.

Argues that the transformation of chemistry into an industrialized big science was a result, at least in part, to an organized use of popularization. Demonstrates the role of World War I in accelerating the process.

144. Rydell, Robert W. *All the World's a Fair: Visions of Empire at American International Expositions, 1876-1916.* Chicago: University of Chicago Press, 1984. x + 328 pp. Bibliography, Index.

Documents how racism and empire served as major themes at the dozen American international expositions during this period. Discusses the scientific underpinning provided racist attitudes through exhibitions developed by Smithsonian Institution anthropologists and utilizing Smithsonian artifacts.

145. Rydell, Robert W. "The Fan Dance of Science: American World's Fairs in the Great Depression." *Isis*, 76 (1985): 525-542.

Utilizes the world fairs of Chicago and New York to demonstrate how scientists during the interwar period attempted to popularize science, create a new American culture which would inculcate scientific values, and validate the hegemony of the corporate state during the interwar period. Credits the National Research Council with taking the leadership role.

PSEUDOSCIENCE

146. Bauer, Henry H. *Beyond Velikovsky: The History of a Public Controversy.* Urbana: University of Illinois Press, 1984. xiii + 354 pp. Index, Bibliography.

Views the Velikovsky affair as an exemplar of a public controversy over scientific matters. Chronicles the events. Demonstrates that Immanuel Velikovsky was a pseudo-scientist. Concludes that Velikovsky's critics committed the same sort of blunders that he did, relying on ad hominem attacks and misrepresentations.

147. Fuller, Robert C. *Mesmerism and the American Cure of Souls.* Philadelphia: University of Pennsylvania Press, 1982. xvi + 227 pp. Bibliography, Index.

Utilizes the history of mesmerism to throw light on the American use of psychological ideas. Sees mesmerism as an effort to account for the inner life of humans in strictly scientific terms. Concludes that mesmerism evolved in the United States into a popular psychology with quasi-religious functions. Identifies Mental Science (Mind Cure) as the post-bellum child of mesmerism.

148. Molella, Arthur P. "At the Edge of Science: Joseph Henry, 'Visionary Theorizers,' and the Smithsonian Institution." *Annals of Science,* 41 (1984): 445-461.

Provides examples of Henry's dealings with unorthodox science at a time when he was a spokesman for professionalism and an advocate of an orthodoxy in science.

149. Sollors, Werner. "Dr. Franklin's Celestial Telegraph, or Indian Blessings to Gas-Lit American Drawing Rooms." *American Quarterly,* 35 (1983): 459-480.

Challenges Van Wyck Brooks's thesis that American culture was divided between visionaries and practical men. Claims that Americans combined the visionary with the practical, and the occult with the scientific.

150. Wrobel, Arthur, editor. *Pseudo-Science and Society in Nineteenth-Century America.* Lexington: The University Press of Kentucky, 1987. vii+245 pp. Index.

Contains papers on phrenology, mesmerism, spiritualism, hydropathy, and homoeopathy. Argues that the pseudo-sciences shared assumptions and methodologies. Describes the pseudo-sciences as rationalistic, egalitarian, and utilitarian.

RELIGION AND SCIENCE

151. Carozzi, Marguerite. "Reaction of British Colonies in America to the 1755 Lisbon Earthquake--A Comparison to the European Response." *Earth Sciences History,* 2 (1983): 17-27.

Contends that Europeans evoked both theological and natural explanations, while American commentators agreed that it was caused by God's will manifested through natural causes.

152. Davis, Dennis Royal. "Presbyterian Attitudes toward Science and the Coming of Darwinism in America, 1859 to 1929." Ph.D. dissertation, University of Illinois at Urbana-Champaign, 1980.

Divides the response into four stages: initially, there were arguments defending the anthrocentric universe; then came two stages of increasing acceptance or accommodation, even among conservatives; the last stage was in response to the rise of fundamentalism. Argues that until the 1920s, opposition to Darwin was based on older scientific frames of reference, not the Bible. Finds no evidence of warfare between science and religion in the Presbyterian Church until the same decade.

153. Ellis, William E. "Evolution, Fundamentalism, and the Historians: A Historiographical Review." *The Historian*, 44 (1981): 15-35.

Finds that there is little agreement over the definition of fundamentalism. Divides the fundamentalist historiography into three schools: participants and observers of the conflict in the 1920s, who were influential into the 1950s; liberal scholars of the 1950s and 1960s; and revisionists, who began to be important in the mid-sixties. Believes that fundamentalists focused on evolution in the 1920s because it was an issue lay persons could identify with and could be attacked outside the church.

154. Gillespie, Neal C. "Preparing for Darwin: Conchology and Natural Theology in Anglo-American Natural History." *Studies in the History of Biology*, 7 (1984): 93-145.

Views the main issues in conchology as the definition of a species and the geographical distribution of a species. Argues that the importance of natural theology in understanding the practice of Anglo-American natural history has been exaggerated. Sees natural theology as solving philosophical and religious problems, not scientific ones. Concludes that natural theology obscured the increasingly positive practice of science during the years prior to Darwin, a practice which enabled naturalists to adopt evolution relatively easily.

155. Greek, Cecil Eugene. "The Religious Roots of American Sociology." Ph.D. dissertation, New School for Social Research, 1983.

Argues that the roots of American sociology can be found in the Protestant Social Gospel Movement of the late nineteenth century, whose leaders were looking for an ameliorative scientific technique to help rid the country of social evils. Claims that sociology has never freed itself from the influence of its religious roots.

156. Larson, Edward J. *Trial and Error: The American Controversy over Creation and Evolution.* New York: Oxford University Press, 1985. Revised edition, 1989. ix + 243 pp. Appendix, Bibliography, Index.

Traces the legal history of efforts to outlaw the teaching of evolution through the 1987 Supreme Court ruling against the Louisiana statute. Links the rise of concern in the early twentieth century to the increase in the number of students attending high school. Sees the legal actions as attempts to reconcile publicly supported science teaching in the high schools with popular opinion on evolution and creation. Views the legislatures and judges as avoiding decisions on the scientific virtues of either evolution or creation; by the 1970s the courts held that the scientific community should be the judges of scientific content. Concludes that the basic issues have not changed since the 1920s.

157. Livingstone, David N. *Nathaniel Southgate Shaler and the Culture of American Science.* Tuscaloosa: University of Alabama Press, 1987. xiv + 395 pp. Appendix, Bibliography, Index.

Presents this book as a response to changes in historiography, especially the efforts to place scientific ideas in their social context. Contends that historians of geography have not responded to such changes in historiography. Claims that during the nineteenth century, geography became the study of human interaction with the environment, as part of the ongoing debate on man's place in the natural order. Sees Shaler as the necessary synthesizer for geography of scientific and religious ideas in the wake of Darwin. Portrays him as a neo-Lamarckian with a "strongly environmental bias," who saw society and nature in a dynamic relationship. Calls him a conservationist rather than a preservationist.

158. Marty, Martin E. "The Impact of Technology on American Religion." *Technology, the Economy, and Society: The American Experience* (item 047), pp. 278-287.

Concludes that technology makes it more difficult for religious institutions to segregate themselves from cultural shifts.

159. Morris, Henry M. *A History of Modern Creationism.* San Diego: Master Book Publishers, 1984. 382 pp. Appendices, Indexes.

Offers an account of the rise of creationism by one of the leading scientific creationists. Illustrates how creationists can develop scientific credibility in experimental fields. Credits *The Genesis Flood* by the author and John C. Whitcomb, Jr. as the catalyst for the rise of modern creationism in the United States.

160. Nelkin, Dorothy. *The Creation Controversy: Science or Scripture in the Schools.* New York: W.W. Norton and Company, 1982. 242 pp. Appendices, Index.

Presents an historical overview. Sees post-World War II creationists as a demonstration of public disillusionment with science and technology as solutions to problems, and an expression of resistance to remote bureaucracies, dominated by experts, who fail to respond to the priorities and needs of the public.

161. Numbers, Ronald L. "Creationism in Twentieth-Century America." *Science,* 218 (1982): 538-544.

Studies the shift of special creationists from a biblical defense of their convictions to one based on science. Notes the change from interpreting the days of creation as symbolic to six literal days and a universal flood.

162. Numbers, Ronald L. "Science and Religion." *Osiris,* 2nd ser., 1 (1985): 59-80.

Blames the influence of Andrew D. White and John W. Draper for the domination of military metaphors in the literature. Finds no evidence of a serious conflict between science and religion. Argues that advocates of the warfare thesis have distorted the complex relationship between the scientist and the theologian.

163. Stephens, Lester D. *Joseph LeConte: Gentle Prophet of Evolution.* Baton Rouge: Louisiana State University Press, 1982. xix + 340 pp. Bibliography, Index.

Traces the life and thought of a geologist who attempted to reconcile evolution and Christianity. Provides insights into the scientific communities of the antebellum South and post-bellum California. Describes LeConte's research in optics and geology.

164. Zanine, Louis J. "From Mechanism to Mysticism: Theodore Dreiser and the Religion of Science." Ph.D. dissertation, University of Pennsylvania, 1981.

    Compares Dreiser's interest in science to a religious quest. Believes he was seeking in science answers to questions about human destiny and purpose which had been traditionally supplied by religion. Discusses the impact of the writings of Herbert Spencer, Charles Darwin, and Jacques Loeb on Dreiser. Concludes that Dreiser's "mechanistic pantheism" was a variation of the eighteenth-century natural theological argument from design.

SCIENCE-TECHNOLOGY RELATIONSHIP

165. Carlson, W. Bernard. "Invention, Science, and Business: The Professional Career of Elihu Thomson, 1870-1900." Ph.D. dissertation, University of Pennsylvania, 1984.

    Attributes Thomson's success as an inventor within a business environment to his ability to interact positively with other managers and engineers. Argues that science supplied the early electrical industry with both theories and "craft" knowledge. Utilizes the concept of technological style to describe Thomson's work. Defines technological style in terms of design motifs, intellectual propensities, marketing perceptions, and organizational abilities.

    Published as *Innovation as a Social Process: Elihu Thomson and the Rise of General Electric, 1870-1900.* Cambridge: Cambridge University Press, 1991. xxii+377 pp.

166. Carroll, P. Thomas. "American Science Transformed." *American Scientist*, 74 (1986): 466-485.

    Focuses on the role of industrialization in the development of American science during the late nineteenth century. Argues that American superiority in precision machinery gave American scientists an edge in the construction of the apparatus that provided the precise measurements and empirical determinations synonymous with American research.

167. Hounshell, David A. "Edison and the Pure Science Ideal in 19th-Century America." *Science,* 207 (1980): 612-617.

      Identifies Edison's claim to be a scientist and the acclaim he received from scientists as the catalysts for Henry A. Rowland's 1883 "plea for pure science."

168. Vincenti, Walter G. "Technological Knowledge without Science: The Innovation of Flush Riveting in American Airplanes, ca. 1930-ca. 1950." *Technology and Culture,* 25 (1984): 540-576.

      Provides schema of technological knowledge. Distinguishes between "descriptive," "prescriptive," and "tacit" knowledge. Calls for attention to techniques of production.

169. Wise, George. "Science and Technology." *Osiris,* 2nd ser., 1 (1985): 229-246.

      Assesses the relationship between science and technology. Sees science policy makers as viewing science and technology as parts of an assembly line with pure science at one end and innovations at the other. Argues that historians reject this view but have not developed an alternative model; instead, they use metaphors. Suggests components of possible models.

CHAPTER III: SCIENCE, TECHNOLOGY, AND GOVERNMENT

GENERAL

170. Armstrong, David A. *Bullets and Bureaucrats: The Machine Gun and the United States Army, 1861-1916.* Westport, Conn.: Greenwood Press, 1982. xv+239 pp. Bibliography, Index.

    Argues that the United States Army failed to exploit the machine gun properly because its bureaucracy had misplaced priorities and was overly conservative. Traces the technological evolution of the machine gun.

171. Bates, Charles C., and John F. Fuller. *America's Weather Warriors, 1814-1985.* College Station: Texas A&M Press, 1986. xxiv+360 pp. Appendices, Bibliography, Index.

    Concentrates on World War II and the Vietnam War. Considers personnel issues, such as the role of women and minorities. Documents the continuing vulnerability of modern, high-technology weapon systems to weather.

172. Giebelhaus, August W. "American Energy Policy." *Perspectives on Public History.* Edited by Brian Greenberg. Wilmington: Hagley Museum and Library, 1985, pp. 7-16.

    Provides an overview of American energy policy toward petroleum, natural gas, and coal. Shows that policies arose in response to crises, rather than as a result of long-range planning. Suggests connections between historical research and policy formation.

173. Nelson, Clifford M. "Paleontology in the United States Federal
       Service, 1804-1904." *Earth Science History,* 1 (1982): 48-
       57.

          Provides a very concise history. Divides the activities
     into three stages: pre-1865 was a period of discontinuous
     contracts; 1865-1879 was marked by collecting as part of the
     federal surveys; from 1879, federal research was conducted by
     the United States Geological Survey.

174. Noble, David F. "Command Performance: A Perspective on the
       Social and Economic Consequences of Military Enter-
       prise." *Military Enterprise and Technological Change:
       Perspectives on the American Experience* (item 176), pp.
       329-346.

          Charges that military enterprise is biased against small
     producers and working people. Identifies the military with high-
     cost, high-performance technology, management by command,
     and large, corporate, capital-intensive manufacturing.

175. Sherwood, Morgan. "The Origins and Development of the
       American Patent System." *American Scientist,* 71 (1983):
       500-506.

          Contends that American society has recognized four
     different rights connected with patents, with the relative
     significance changing over time.

176. Smith, Merritt Roe, editor. *Military Enterprise and Technological
       Change: Perspectives on the American Experience.*
       Cambridge: MIT Press, 1985. x+391 pp. Index.

          Identifies five themes: design and dissemination of new
     technologies; management; testing, instrumentation, and quality
     control; uniformity and order; and innovation and social process.
     Includes items 013, 174, 186, 189, 193, 198, 230, 520, 607.

177. Thibodeau, Sharon Gibbs. "Science in the Federal Government."
       Osiris, 2nd ser., 1 (1985): 81-96.

          Concentrates on publications post-Dupree (Item I:294);
     i.e., items published after 1957 and focusing on post-1940
     events. Finds that this literature has tended to be on specific
     institutions, projects, or developments within disciplines.
     Identifies three different perspectives--the professional historian,
     the reflective participant, and the critical observer-advocate.

Provides a short review of primary sources available at the National Archives.

NINETEENTH CENTURY

178. Aldrich, Michele L. "Charles Thomas Jackson's Geological Surveys in New England, 1836-1844." *Northeastern Geology,* 3 (1981): 5-10.

    Defends the use of a travelogue style in Jackson's reports as both practical and in concert with Jackson's philosophy that his contribution to science should be accurate descriptions.

179. Aldrich, Michele L., and Alan E. Leviton. "James Hall and the New York Survey." *Earth Sciences History,* 6 (1987): 24-33.

    Surveys Hall's life and work. Finds Hall's institutional base in New York to be unstable on occasion, but very important for his scientific career.

180. Block, Robert H. "The Whitney Survey of California, 1860-1874: A Study of Environmental Science and Exploration." Ph.D. dissertation, University of California at Los Angeles, 1982.

    Sees the California State Geological Survey under Josiah Dwight Whitney as a link between earlier Humboldtean-style exploration and the later more specialized scientific surveys. Treats Whitney's survey as a geographical operation, using Reingold's definition (see Item I:411). Contends the Survey provided part of the foundation for twentieth-century environmental sciences.

181. Fakundiny, Robert H., and Ellis L. Yochelson, editors. "Special James Hall Issue." *Earth Sciences History,* 6, no. 1 (1987).

    Contains sixteen papers focusing on Hall, the history of the New York State Geological Survey, and the relationship of the New York survey with other state and national surveys. Includes items 88, 179, 283.

182. Hafertepe, Kenneth. *America's Castle: The Evolution of the Smithsonian Building and Its Institution, 1840-1878.* Washington, D.C.: Smithsonian Institution Press, 1984. xxiii + 180 pp. Bibliography, Index.

Discusses the conflicts over the design, construction, and use of the Smithsonian Institution's "Castle" during the period that Joseph Henry was Secretary. Assesses the role the form of the building had in defining the functions of the Smithsonian.

183. Jordan, William M., and Norman A. Pierce. "J. Peter Lesley and the Second Geological Survey of Pennsylvania." *Northeastern Geology,* 3 (1981): 75-85.

Argues that Lesley's experiences as an assistant to Henry Darwin Rogers on the First Pennsylvania Geological Survey were critical influences on his organization of the Second Survey and his relationship with assistants. Summarizes Lesley's contributions to geology.

184. Millbrooke, Ann. "South Carolina State Geological Surveys of the Nineteenth Century." *The Geological Science in the Antebellum South* (item 279), pp. 26-38.

Contends that these surveys were established as limited agencies with practical goals. Argues that this was typical of the institutions and uses of science in the South.

185. Millbrooke, Ann. "State Geological Surveys of the Nineteenth Century." Ph.D. dissertation, University of Pennsylvania, 1981.

Provides four case studies: South Carolina, Pennsylvania, Illinois, and California. Views the surveys as responses to special interest groups within states. Finds that by the end of the century state surveys were permanent bureaus of technical information with responsibility for both scientific research and public service.

186. O'Connell, Charles F., Jr. "The Corps of Engineers and the Rise of Modern Management, 1827-1856." *Military Enterprise and Technological Change: Perspectives on the American Experience* (item 176), pp. 87-116.

Challenges Alfred D. Chandler, Jr.'s thesis that the railroads were responsible for creating modern industrial management. Credits the Corps of Engineers' management techniques with supplying the model for railroad management.

187. Preston, Daniel. "The Administration and Reform of the U. S. Patent Office, 1790-1836." *Journal of the Early Republic,* 5 (1985): 331-353.

Argues that the 1836 reform was the result of new recognition of the significance of patents and inventions for the American economy, the Jacksonian philosophy toward government, and the breakdown in the efficiency of the patent office after the death of William Thorton.

188. Reuss, Martin. "Andrew A. Humphreys and the Development of Hydraulic Engineering: Politics and Technology in the Army Corps of Engineers, 1850-1950." *Technology and Culture,* 26 (1985): 1-33.

Argues that Humphreys oversaw and, in part, caused the decline in the reputation of the Corps of Engineers, through his defense of his theory of using only levees for flood control. Documents how a bureaucracy can elevate a theory into a dogma.

189. Smith, Merritt Roe. "Army Ordnance and the 'American system' of Manufacturing, 1815-1861." *Military Enterprise and Technological Change: Perspectives on the American Experience* (item 176), pp. 39-86.

Argues that private entrepreneurs have been given too much credit.

190. Smith, Merritt Roe. "Military Entrepreneurship." *Yankee Enterprise: The Rise of the American System of Manufactures* (item 609), pp. 63-102.

Argues that the Ordnance Department facilitated the rise of the American System of Manufacturing through its efforts to secure uniformity in weapons in the wake of the War of 1812. Traces the diffusion of skills and techniques from government armories to civilian machine tool shops.

191. Stanley, Autumn, "The Patent Office Clerk as Conjurer: The Vanishing Lady Trick in a Nineteenth-Century Source," *Women, Work, and Technology: Transformations.* Edited by Barbara Drygulski Wright, et al. Ann Arbor: University of Michigan Press, 1987, pp. 118-136.

Demonstrates that the official list of women to whom American patents were issued in the nineteenth century is in error. Argues that this is especially true for mechanical inventions. Believes the list reflects and reinforces stereotypes.

TWENTIETH CENTURY

192. Allison, David Kite. *New Eye for the Navy: The Origin of Radar
      at the Naval Research Laboratory.* Washington, D.C.:
      Naval Research Laboratory, 1981. xi+228 pp. Appen-
      dices, Bibliography, Index.

      Stresses an institutional approach to the history of radar
      at the Naval Research Laboratory. Argues that radar
      demonstrated the value of a mission-oriented research and
      development facility. This is a revision of Item I:280.

193. Allison, David Kite. "U. S. Navy Research and Development
      since World War II." *Military Enterprise and Tech-
      nological Change: Perspectives on the American
      Experience* (item 176), pp. 289-328.

      Studies naval bureaucracy and organization. Provides two
      case studies--Sidewinder and the Navy Tactical Data System.

194. Becker, John V. *The High-Speed Frontier: Case Histories of
      Four NACA Programs, 1920-1950.* Washington, D.C.:
      National Aeronautics and Space Administration, 1980.
      viii+190 pp. Appendix, Bibliography, Index.

      Mixes personal reminiscences with technical commentary
      on the high-speed airfoil program; transonic wind tunnel
      development; high-speed propeller program; and high-speed
      cowlings, air inlets and outlets, and internal-flow systems.
      Utilizes interviews with twenty-two participants in the research.
      Provides insights into the personalities involved.

195. Bernstein, Barton J. "'In the Matter of J. Robert Oppenheimer'."
      Historical Studies in the Physical Sciences, 12 (1982):
      195-252.

      Concludes that Oppenheimer's actions were the result of
      his inability to choose between an older morality of individual
      conscience and the new one of higher loyalty to the state. Finds
      that the Oppenheimer case did not permanently damage the
      relationship between scientists and the United States government.

196. Bilstein, Roger E. *Stages to Saturn: A Technological History of
      the Apollo/Saturn Launch Vehicles.* Washington, D.C.:
      National Aeronautics and Space Administration, 1980.
      xx+511 pp. Appendices, Bibliography, Index.

Explains how the Saturn Launch Vehicles were constructed and how they operated. Organizes the information by topic. Includes an abundance of technical detail. Provides insight into the solving of design and construction problems in high-technology projects.

197. Bromberg, Joan Lisa. *Fusion: Science, Politics, and the Invention of a New Energy Source*. Cambridge: MIT Press, 1982. xxvi + 344 pp. Appendices, Index.

Studies the research program which arose from intersection of plasma physics with society's desire for a clean, abundant energy source. Focuses on the program strategy through 1978. Uses no classified information. Discovers that the political and social demands placed on the program have continuously changed, as have the rationales. Credits Robert Louis Hirsch, named Director of the Division of Controlled Thermonuclear Research in 1972, with converting fusion research from small to big science and changing the goal from demonstration of feasibility to a reactor.

198. Buck, Peter. "Adjusting to Military Life: The Social Sciences Go to War, 1941-1950." *Military Enterprise and Technological Change: Perspectives on the American Experience* (item 176), pp. 203-252.

Examines the sociological research conducted by the Army to help civilians adjust to military life. Argues that their military experience influenced the founders of the postwar Department of Social Relations at Harvard University to develop a sociology stressing harmony, consensus, and order.

199. Compton, W. David and Charles D. Benson. *Living and Working in Space: A History of Skylab*. Washington, D.C.: National Aeronautics and Space Administration, 1983. xiii + 449 pp. Appendices, Index.

Provides a full, detailed history of the planning, development, and missions of this effort to continue American manned spaceflight beyond Apollo. Places the controversies and problems of the Skylab program in context. Fails to provide an analysis of the scientific contributions of the program.

200. Cravens, Hamilton. "Applied Science and Public Policy: The Ohio Bureau of Juvenile Research and the Problem of Juvenile Delinquency, 1913-1930." *Psychological Testing and American Society, 1890-1930* (item 470), pp. 158-194.

Traces the progression of Henry H. Goddard's theory from a crude hereditarian determinism to a more complex view which allowed for environment and training to overcome the innate low intelligence which led to crime. Demonstrates how the lack of political skills could defeat efforts to apply scientific expertise.

201. DeVorkin, David H. "Organizing for Space Research: The V-2 Rocket Panel." *Historical Studies in the Physical Sciences,* 18 (1987): 1-24.

Provides an account of activities from 1946 to 1952. Sees the V-2 Panel as a substitute for a disciplinary infrastructure for space science. Accepts James A. Van Allen's view that the legacy of the V-2 Panel era was a body of scientists trained in the basic techniques of space science and in the skills necessary to utilize a military program for patronage.

202. Dupree, A. Hunter. "The History of the Exploration of Space: From Official Historians to Contributors to Historical Literature." *Public Historian,* 8 (1986): 121-128.

Essay review. Defends history contracted by the National Aeronautics and Space Administration. Argues that the most recent official histories are very close to the quality of academic history.

203. England, J. Merton. *A Patron for Pure Science.* Volume I: *The National Science Foundation's Formative Years, 1945-1957.* Washington, D.C.: National Science Foundation, 1983. x + 443 pp. Appendices, Bibliography, Index.

Provides a detailed legislative and administrative history of the National Science Foundation through Sputnik. Includes detailed analysis of regulatory biology in the 1950s as an example of the pre-1954 support for smaller scientific projects. Argues that after 1954, interest shifted to big science projects. Fails to place the history of the National Science Foundation into the larger political and scientific context.

204. Ezell, Edward Clinton, and Linda Neumen Ezell. *On Mars: Exploration of the Red Planet, 1958-1978.* Washington, D.C.: National Aeronautics and Space Administration, 1984. xvi+535 pp. Appendices, Bibliography, Index.

Focuses on Project Viking. Documents the adverse impact of manned spaceflight on the funding of space science.

205. Forman, Paul. "Behind Quantum Electronics: National Security as Basis for Physical Research in the United States, 1940-1960." *Historical Studies in the Physical and Biological Sciences,* 18 (1987): 149-229.

Argues that American physics underwent a qualitative change during this period, as a result of its involvement in national security research. Believes that physicists lost control of their discipline, although they maintained the illusion of autonomy.

206. Genuth, Joel. "Groping Towards Science Policy in the United States in the 1930s." *Minerva,* 25 (1987): 238-268.

Evaluates the efforts by the Science Advisory Board and the National Resources Planning Board to develop an agenda for American science policy during the 1930s. Contrasts the Science Advisory Board's vision of academic scientists advising civil servants at the government science bureaus with the National Resources Planning Board's concept of a group of scientists advising the president. Concludes that neither board could gain a consensus among the scientific community.

207. Gerstein, Dean R. "Social Science as a National Resource 1948 and 1982." *The Nationalization of the Social Sciences* (item 217), pp. 247-264.

Contrasts the success of the National Academy of Sciences/National Research Council 1982 report, *Behavioral and Social Science Research: A National Resource,* with the failure of Talcott Parsons's "Social Science: A Basic National Resource." Finds that Parsons stressed the unity of the social sciences and basic research; the 1982 report stressed disciplines and did not celebrate pure research. Concludes that the use of a committee instead of an individual author helped.

208. Gillmor, C. Stewart. "Federal Funding and Knowledge Growth in Ionospheric Physics, 1945-81." *Social Studies of Science,* 16 (1986): 105-133.

Focuses on university research. Concludes that the data reveals no direct links between individual funding programs and intellectual changes in the field. Suggests that the situation may be too complex for clear-cut relationships to emerge.

209. Graybar, Lloyd J. "The 1946 Atomic Bomb Tests: Atomic Diplomacy or Bureaucratic Infighting." *The Journal of American History*, 72 (1986): 888-907.

Concludes that the Bikini Atoll tests were made for military reasons, not to impact upon international diplomacy.

210. Hacker, Barton C. *The Dragon's Tail: Radiation Safety in the Manhattan Project, 1942-1946.* Berkeley: University of California Press, 1987. x+258 pp. Appendix, Bibliography, Index.

Finds that the threshold approach dominated the thought of Manhattan Project programs. Admits that safety was not the highest priority at Los Alamos, but attributes this to wartime contingencies. Concludes that given the circumstances and limited knowledge concerning radiation safety prior to 1942, government practices were reasonable.

211. Hallion, Richard P. *On the Frontier: Flight Research at Dryden, 1946-1981.* Washington D.C.: National Aeronautics and Space Administration, 1984. xix+385 pp. Appendices, Index.

Traces the history of flight research at the Hugh L. Dryden Flight Research Center under the National Advisory Committee for Aeronautics and the National Aeronautics and Space Administration. Identifies its two major areas of contributions as early supersonic flight technology and flight beyond the atmosphere. Provides a balance between institutional and technical history.

212. Hansen, James R. *Engineer in Charge: A History of the Langley Aeronautical Laboratory, 1917-1958.* Washington, D.C.: National Aeronautics and Space Administration, 1987. xxxviii+620 pp. Appendices, Guide to Sources, Index.

Provides a technical history of the National Advisory Committee for Aeronautics. Complements the administrative history by Roland (item 239). Concentrates on the technological achievements of the laboratory, including their contributions to aircraft design. Argues that World War II caused the greatest changes in Langley by expanding and heterogizing the staff and shifting the emphasis from general to specific testing; these changes were not reversed when the war ended. Presents a considerable amount of reference material in the appendices. The guide to NACA historical sources at Langley is very useful.

213. Herken, Gregg. *The Winning Weapon: The Atomic Bomb in the Cold War 1945-1950.* New York: Vantage Books, 1982. xiii + 425 pp. Bibliography, Index.

Assesses the role of the atomic bomb as an instrument of national policy. Documents the Truman Administration's belief that the possession of the bomb would force the Soviets to behave.

214. Jones, Vincent C. *Manhattan: The Army and the Atomic Bomb.* Washington, D.C.: Center of Military History, 1985. xx + 660 pp. Appendix, Bibliography, Index.

Provides an overview of the project to build the atomic bomb from the perspective of the United States Army. Emphasizes the support activities of the Army on behalf of the project. This is an official history.

215. Kargon, Robert, and Elizabeth Hodes. "Karl Compton, Isaiah Bowman, and the Politics of Science in the Great Depression." *Isis,* 76 (1985): 301-318.

Presents Compton as an example of the new type of leader of science who appeared during the period 1930-1945, more receptive to a partnership with government and the need for long-term planning.

216. Klausner, Samuel Z. "The Bid to Nationalize American Social Science." *The Nationalization of the Social Sciences* (item 217), pp. 3-39.

Describes the context of Parsons's essay, which was to be a social science analog to Vannevar Bush's *Science, The Endless Frontier.* Summarizes the criticism of the essay. Concludes that this effort by the Social Science Research Council failed because it was a scholarly council, not a political council.

217. Klausner, Samuel Z., and Victor M. Lidz, editors. *The Nationalization of the Social Sciences.* Philadelphia: University of Pennsylvania Press, 1986. xiv + 296 pp. Indices.

Publishes Talcott Parsons's 1948 manuscript "Social Science: A Basic National Resource." Places the manuscript in context. Reviews the relation between politics and scholarship in the social sciences. Analyzes other aspects of Parsons's theoretical work. Includes items 207, 216, 226, 235, 238, 250, 475.

218. Koppes, Clayton R. *JPL and the American Space Program: A History of the Jet Propulsion Laboratory.* New Haven: Yale University Press, 1982. xiii+299 pp. Indices.

Provides an institutional history--with relatively little technological detail--tracing the evolution of a science and technology research center funded by the federal government at the time the United States was a national security state. Illuminates the relationship among the Jet Propulsion Laboratory, which sought to maintain as much independence as possible, the National Aeronautics and Space Administration, which attempted to treat JPL as a NASA center, and the California Institute of Technology, the parent of JPL.

219. Krammer, Arnold. "Technology Transfer as War Booty: The U.S. Technical Oil Mission to Europe, 1945." *Technology and Culture,* 22 (1981): 68-103.

Views synthetic petroleum as the most representative of the transfers of industrial technology as war booty. Traces American efforts to learn the German technological secrets. Concludes that the United States quickly lost interest after the war because of the availability of Middle East oil.

220. Lambright, W. Henry. *Presidential Management of Science and Technology: The Johnson Presidency.* Austin: University of Texas Press, 1985. xii+224 pp. Index.

Utilizes twenty-four case studies to explore the presidential management of science and technology policy. Focuses on the issues of the dispersion of the science/technology enterprise throughout the government; the sequential nature of presidential decision making; the need to integrate the technical, political, administrative, and budgetary aspects of decision making; and coherence of the policies. Argues that the president influences, but does not dictate, policy direction. Characterizes Johnson as competitive, nationalistic, and pragmatic. Concludes that Johnson had an ad hoc reactive policy in which he was a significant decision maker.

221. Lear, Linda J. "The Boulder Canyon Project: A Re-examination of Federal Resource Management." *Energy in American History* (item 050), pp. 329-337.

   Views the Boulder Dam as a major turning point in natural resource management by the federal government. Contends that it established the principle that hydroelectric power was an essential element of reclamation projects, providing precedent for the Tennessee Valley Authority.

222. Leslie, Stuart W. "Playing the Education Game to Win: The Military and Interdisciplinary Research at Stanford." *Historical Studies in the Physical and Biological Sciences,* 18 (1987): 55-88.

   Discusses the role of the military in the financial and intellectual development of science and engineering programs at Stanford after World War II. Credits Stanford's success to its practice of utilizing the military funds in interdisciplinary laboratories. Sees these laboratories as obscuring the boundaries between disciplines, between science and engineering, and between basic and applied science.

223. Levine, Arnold S. *Managing NASA in the Apollo Era.* Washington, D.C.: National Aeronautics and Space Administration, 1982. xxi+342. Appendices, Index.

   Views administrative history as an account of the interaction between an organization and its environment. Sees NASA as combining centralized planning with decentralized project execution. Attributes NASA's accomplishments to its administrative flexibility, the political ability of senior management, the delegation of program management to field offices, and the timeliness of its decisions.

224. Logsdon, John. "Opportunities for Policy Historians: The Evolution of the U.S. Civilian Space Program." *A Spacefaring People: Perspective on Early Spaceflight.* Edited by Alex Roland. Washington, D.C.: National Aeronautics and Space Administration, 1985, pp. 81-107.

   Identifies six principles of American space policy.

225. Lyon, Edwin Austin, II. "New Deal Archaeology in the Southeast: WPA, TVA, NPS, 1934-1942." Ph.D. dissertation, Louisiana State University and Agricultural and Mechanical College, 1982.

Looks at Works Progress Administration, Civil Works Administration, Tennessee Valley Authority, and National Park Service archaeological projects. Notes that the Smithsonian Institution supplied technical advice. Concludes that the programs resulted in a broad regional interest in the prehistory of the Southeast.

226. Lyons, Gene L. "The Many Faces of Social Science." *The Nationalization of the Social Sciences* (item 217), pp. 197-208.

Maintains that the publication of Talcott Parsons's 1948 manuscript would not have changed history. Describes the National Science Foundation model of social sciences as emphasizing either quantification or linkages to other sciences.

227. McDougall, Walter A. . . . *the Heavens and the Earth: A Political History of the Space Age.* New York: Basic Books, 1985. xviii+555 pp. Appendix, Index.

Intertwines the history of the technology with the political events which both drove and were driven by the technology. Alternatively examines the American and Russian programs in detail through 1964. Traces how these space programs have led to the massive inundation of technological potential into modern society and the evolution of this society into technocracy. Views the space race as a threat to the higher values in society.

228. Mack, Pamela Etter. "The Politics of Technological Change: A History of Landsat." Ph.D. dissertation, University of Pennsylvania, 1983.

Presents the Landsat project--satellite photography for use by resource managers--as a case study of the history of technological change in a government context and the shaping of technology by politics. Concludes that the cause of the majority of the problems was the lack of adequate coordination between the developers of the new technology and the users of the data.

Published as *Viewing the Earth: The Social Construction of the Landsat Satellite System.* Cambridge, Mass.: MIT Press, 1990. xii+270 pp. Bibliography, Index.

229. Mirabito, Michael M. *The Exploration of Outer Space with Cameras: A History of the NASA Unmanned Spacecraft Missions.* Jefferson, N.C.: McFarland, 1983. vi + 170 pp. Bibliography, Appendix, Index.

Reviews the missions. Integrates information provided by the various spacecraft for each of the objects in the solar system. Discusses camera design, imaging, and optics. Serves as a basic guide.

230. Misa, Thomas J. "Military Needs, Commercial Realities, and the Development of the Transistor, 1948-1958." *Military Enterprise and Technological Change: Perspectives on the American Experience* (item 176), pp. 253-287.

Shows how the Army Signal Corps served as an entrepreneur in advancing solid-state physics.

231. Muenger, Elizabeth A. *Searching the Horizon: A History of Ames Research Center, 1940-1976.* Washington, D.C.: National Aeronautics and Space Administration, 1985. xiii + 299 pp. Appendices, Bibliography, Index.

Offers a chronological account. Focuses on and provides excellent insights into the administration and management of the center. This is not a useful source for detailed information on the technological achievements of the center.

232. Needell, Allan A. "Lloyd Berkner, Merle Tuve, and the Federal Role in Radio Astronomy." *Osiris,* 2nd ser., 3 (1987): 261-288.

Contrasts Tuve's opinion that the National Science Foundation should support researchers, not facilities, with Berkner's desire to have the NSF supply funds to the Associated Universities, Incorporated, for the Green Bank National Radio Observatory.

233. Needell, Allan A. "Nuclear Reactors and the Founding of Brookhaven National Laboratory." *Historical Studies in the Physical Sciences,* 14 (1983): 93-122.

Presents Brookhaven as an example of scientists underestimating the difficulties of managing large research and development efforts. Argues that Brookhaven displayed the tensions between concepts of good science and good management and scientific independence and public accountability. Sees Brookhaven as ultimately serving as a model for large, government-sponsored scientific institutions.

234. Needell, Allan A. "Preparing for the Space Age: University-based Research, 1946-1957." *Historical Studies in the Physical Sciences,* 18 (1987): 89-109.

Argues that the research on cosmic rays and the electromagnetic properties of near-space earth during the late 1940s and early 1950s prepared scientists for the application of space technology. Finds that two characteristics of university-based space science--dependence on outside agencies for funding and technical support, and the compromise between the criteria of the supporting agencies and academic practices--are evident in pre-1957 research in the physical sciences.

235. Prewitt, Kenneth. "Federal Funding for Social Science." *The Nationalization of the Social Sciences* (item 217), pp. 227-238.

Points out that funds provided by mission agencies because of policy considerations were more important to the social sciences, in terms of total dollars, than the funds coming from the National Science Foundation.

236. Rabbitt, Mary C. *Minerals, Lands, and Geology for the Common Defense and General Welfare.* Volume III: *1904-1939.* Washington, D.C.: Government Printing Office, 1986. x + 479 pp. Bibliography, Indices.

Provides a chronological history of the United States Geological Survey during a period when the Survey was transformed first from a research organization to an institution for applied science, and then back to a research organization. Focuses on the issue of conservation. This is a continuation of Items I:346 and I:347.

237. Reingold, Nathan. "Vannevar Bush's New Deal for Research: or the Triumph of the Old Order." *Historical Studies in the Physical and Biological Sciences,* 17 (1987): 299-344.

Pictures Bush as a political conservative opposed to government science and in favor of collegial governance. Presents the National Advisory Board for Aeronautics as the model for Bush's National Science Foundation proposal. Argues that Bush saw the National Science Foundation as the federal entity, not simply a funnel to provide federal dollars to the universities. Views the debate over whether the President or the National Science Board appointed the Director of the National Science Foundation as unimportant.

238. Riecken, Henry W. "Underdogging: The Early Career of the Social Sciences in the NSF." *The Nationalization of the Social Sciences* (item 217), pp. 209-225.

Credits the growth of support by the National Science Foundation for the social sciences during the years 1954-1968 to external support by respected advisors, skilled administrators within the NSF, and the lack of serious political errors by grantees, rather than any scientific accomplishments. Sees the NSF's emphasis on positivistic, quantitative research and the methodological similarities of the social and natural sciences as a response to the nervousness of public officials about the social sciences.

239. Roland, Alex. *Model Research: The National Advisory Committee for Aeronautics, 1915-1958.* Washington, D.C.: National Aeronautics and Space Administration, 1985. 2 volumes. xxix+769+25 pp. Bibliography, Appendices, Index.

Provides a political and institutional history from the perspective of NACA headquarters. Argues that NACA's lack of a political base left it insecure and afraid for its survival. Attacks NACA for valuing loyalty and teamwork more than brilliance. Contends that the wind tunnel dominated its research program. Serves as a guide to the archival records of NACA. Over half the book is in the form of appendices.

240. Rossiter, Margaret W. "Science and Public Policy since World War II." *Osiris,* 2nd ser. 1 (1985): 273-294.

Organizes the bibliographic essay according to eight themes, including science education, the space program, and safety and environmental issues. Observes that science has been involved in many of the central issues of public policy in the years after 1945.

241. Rowan, Milton. "Politics and Pure Research: The Origins of the National Science Foundation." Ph.D. dissertation, Miami University, 1985.

Traces the debate over the nature of the National Science Foundation. Sees the NSF as neither a coordinator of government research, nor as a vehicle for reform as envisioned by Harley

Kilgore, but as a fulfillment of Vannevar Bush's objective of government subsidy of basic research and scientific training without disturbing the established structure of American science.

242. Seidel, Robert W. "Accelerating Science: The Postwar Transformation of the Lawrence Radiation Laboratory." *Historical Studies in the Physical Sciences,* 13 (1983): 375-400.

Discusses Lawrence's ability to cultivate both General Leslie R. Groves, head of the Manhattan Engineer District, and the leadership of the Atomic Energy Commission, and to sell his research program for large accelerators.

243. Seidel, Robert W. "From Glow to Flow: A History of Military Laser Research and Development." *Historical Studies in the Physical and Biological Sciences,* 18 (1987): 111-147.

Focuses on the conversion of military laser technology from solid state to flowing gas lasers. Argues that institutional self-interest, mission conception, and the commitments made by developers to the Department of Defense have resulted in a premature scaling up of the devices.

244. Seidel, Robert W. "A Home for Big Science: The Atomic Energy Commission's Laboratory System." *Historical Studies in the Physical and Biological Sciences,* 6 (1986): 35-75.

Focuses upon its patronage system. Views the laboratories of the Atomic Energy Commission as the new homes for Big Science in post-war America. Finds that the price of support for scientists at these laboratories was the acceptance of security restrictions and military applications of science.

245. Stine, Jeffrey K. *A History of Science Policy in the United States, 1940-1985.* Report Prepared for the Task Force on Science Policy of the House Committee on Science and Technology. 99th Congress, 2d sess. Background Report No. 1. viii + 120 pp. Appendix.

Summarizes key events.

246. Stine, Jeffrey K., and Gregory A Good. "Government Funding of Scientific Instrumentation: A Review of U.S. Policy Debates Since World War II." *Science, Technology and Human Values,* Issue 3 (Summer 1986): 34-46.

Observes that apprehension about the adequacy of scientific instrumentation is not new. Finds that the issue surfaces as a major policy concern after prolonged periods of underinvestment in instrumentation.

247. Tatarewicz, Joseph N. "Federal Funding and Planetary Astronomy, 1950-75: A Case Study." *Social Studies of Science,* 6 (1986): 79-103.

Argues that the National Aeronautics and Space Administration attempted to strengthen ground-based planetary astronomy to assist the space program. Attempts to measure the increased interest in planetary astronomy in objective ways and to relate it to the infusion of funds by NASA.

248. Trenn, Thaddeus J. *America's Golden Bough: The Science Advisory Intertwist.* Cambridge, Mass.: Oelgeschlager, Gunn and Hain, 1983. xxviii+307 pp. Appendices, Bibliography, Index.

Utilizes the evolution of science advice to the presidency as the framework for a study of the intertwining of American science and politics from World War II through 1981. Warns that this advice was only one aspect of the total system. Sees the American science advisory system as a pragmatic response to the intertwist, defined more in terms of function and purpose than substance and institutional structure. The writing style is more akin to that of a textbook than a historical analysis.

249. Tuttle, William M., Jr. "The Birth of an Industry: The Synthetic Rubber 'Mess' in World War II." *Technology and Culture,* 22 (1981): 35-67.

Views the mass production of synthetic rubber as enhancing the image of scientists and engineers as problem-solvers. Highlights the conflict between outlooks dominated by wartime exigencies and those worrying about postwar considerations.

250. Wolfe, Dael. "Making a Case for the Social Sciences." *The Nationalization of the Social Sciences* (item 217), pp. 85-96.

Identifies three reasons why the social sciences were not among the disciples that the National Science Foundation was mandated to support in 1950: lack of political influence by social scientists, a skepticism over the scientific nature of the social sciences, and the feeling by many of the participants in the debate over the nature of the NSF that the inclusion of the social sciences was not as important as other issues. Sees the exclusion as a blow to the pride of the social sciences; it had little impact on the amount of funding received.

CHAPTER IV: THE PHYSICAL SCIENCES

ASTRONOMY

251. Crelinsten, Jeffrey. "William Wallace Campbell and the 'Einstein Problem': An Observational Astronomer Confronts the Problem of Relativity." *Historical Studies in the Physical Sciences*, 14 (1983): 1-91.

Finds that Campbell lacked understanding of Relativity Theory. Concludes that American astronomers were inferior in mathematical skills to their British counterparts.

252. DeVorkin, David H. "The Maintenance of a Scientific Institution: Otto Struve, the Yerkes Observatory, and Its Optical Bureau During the Second World War." *Minerva*, 18 (1980): 595-623.

Argues that the Optical Bureau was Struve's response to his fear that the dispersal of the staff would result in the closing of the observatory. He sought a research program which was militarily useful, based on astronomical knowledge, and would keep the staff at the observatory.

253. DeVorkin, David H. and Ralph Kenat. "Quantum Physics and the Stars (II): Henry Norris Russell and the Abundance of the Elements in the Atmosphere of the Sun and Stars." *Journal for the History of Astronomy*, 14 (1983): 180-222.

Traces the steps leading to Russell's 1929 paper, "On the Composition of the Sun's Atmosphere." Concludes that the paper is an example of Russell's ability to integrate evidence and come to a generalization.

254. Dick, Stephen J. "How the U.S. Naval Observatory Began, 1830-65." *Sky with Ocean Joined: Proceedings of the Sesquicentennial Symposia of the U.S. Naval Observatory* (item 255), pp. 167-181.

Outlines the evolution of the Naval Observatory from the establishment of the Depot of Charts and Instruments to the death of James M. Gilliss, the second Superintendent of the Naval Observatory.

255. Dick, Stephen J., and LeRoy E. Doggett, editors. *Sky with Ocean Joined: Proceedings of the Sesquicentennial Symposia of the U.S. Naval Observatory.* Washington, D.C.: U.S. Naval Observatory, 1983. viii + 190 pp.

Contains reviews of contemporary research as well as historical articles. Includes items 254, 261, 265.

256. Hetherington, Norriss S. "Philosophical Values and Observations in Edwin Hubble's Choice of a Model of the Universe." *Historical Studies in the Physical Sciences,* 13 (1982): 41-67.

Argues that Hubble chose general philosophical principles over observational data.

257. Hoyt, William Graves. *Coon Mountain Controversies: Meteor Crater and the Development of Impact Theory.* Tucson: University of Arizona Press, 1987. xii + 443 pp. Index.

Traces the controversy over whether the Coon Mountain Crater (Meteor Crater) in Arizona was the result of a meteor impact. Provides an insightful evaluation of Daniel M. Barringer, a lawyer and engineer, who believed the crater was caused by impact and sought the meteorite mass he thought lay underneath the crater floor. Places the controversy in the context of the debate over the origin of the lunar craters and the more general issue of the role of meteorite impact in the history of the solar system.

258. Hoyt, William Graves. "G. K. Gilbert's Contribution to Selenology." *Journal for the History of Astronomy,* 13 (1982): 155-167.

Argues that Gilbert's impact hypothesis for the origin of lunar craters had little or no impact on either his contemporaries or scientists who later developed impact theory. Rejects the view of Gilbert's influence presented in Item I:457.

259. Kidwell, Peggy A. "Cecilia Payne-Gaposchkin: Astronomy in the Family." *Uneasy Careers and Intimate Lives: Women in Science, 1789-1979.* Edited by Pnina G. Abir-Am and Dorinda Outram. New Brunswick: Rutgers University Press, 1987, pp. 216-238.

Argues that Payne-Gaposchkin's life is fully understandable only in the context of the constraints imposed and the support provided by family life.

260. Lankford, John. "Photography and the Long-focus Visual Refractor: Three American Case Studies, 1885-1914." *Journal for the History of Astronomy,* 14 (1983): 77-91.

Describes the failure to adapt a large refractor for astrophotography at Lick; the successful adaptation at Yerkes; and the decision to build the Thaw Refractor as a photographic instrument, using the successful research program at Yerkes as a model. Concludes astronomers moved very cautiously.

261. Norberg, Arthur L. "Simon Newcomb's Role in the Astronomical Revolution of the Early Nineteen Hundreds." *Sky with Ocean Joined: Proceedings of the Sesquicentennial Symposia of the U.S. Naval Observatory* (item 255), pp. 75-88.

Focuses on Newcomb's revisions of planetary table theory while he was Superintendent of the Nautical Almanac.

262. Osterbrock, Donald E. *James E. Keeler, Pioneer American Astrophysicist and the Early Development of American Astrophysics.* Cambridge: Cambridge University Press, 1984. xii+411 pp. Bibliography, Index.

Provides a well-documented, detailed study of the life and work of the most outstanding American astrophysicist of the late nineteenth century. Discusses Keeler's activities as a staff member and subsequently director of the Allegheny Observatory and Lick Observatory.

263. Osterbrock, Donald E. "The Rise and Fall of Edward S. Holden." *Journal for the History of Astronomy,* 15 (1984): 81-127, 151-176.

Assesses Holden as a good assimilator and organizer of research, but with little personal research skill or creative power. Details his activities as director of Lick Observatory. Documents his personality clashes with his staff.

264. Rothenberg, Marc. "History of Astronomy." *Osiris*, 2nd ser. 1 (1985): 117-131.

Finds very few historians who view this topic as their primary research subject. Concludes that this is a very undeveloped field of investigation.

265. Rothenberg, Marc. "Observers and Theoreticians: Astronomy at the Naval Observatory, 1845-1861." *Sky with Ocean Joined: Proceedings of the Sesquicentennial Symposia of the U.S. Naval Observatory* (item 255), pp. 29-38.

Argues that the astronomical education of the staff of the Naval Observatory was typical of American astronomers as a whole during this period. Contends that the relative low historical visibility of the staff was due to the research programs undertaken by the observatory. Views Matthew Fontaine Maury as a well-meaning outsider who did not understand the American astronomical community.

266. Webb, George Ernest. *Tree Rings and Telescopes: The Scientific Career of A. E. Douglass.* Tucson: University of Arizona Press, 1983. xiii+242 pp. Bibliography, Index.

Examines the professional career and work of Andrew Elliott Douglass in both astronomy and dendrochronology, but presents little insight into his personal life. Demonstrates that the interest in tree rings initially arose because of a fairly common interest among astronomers in the relationships between solar and meteorological activity. Discusses the establishment and construction of the Steward Observatory.

267. Wright, Helen. *James Lick's Monument: The Saga of Captain Richard Floyd and the Building of the Lick Observatory.* Cambridge: Cambridge University Press, 1987. xvi+231 pp. Index.

Documents the work of the overseer of the construction of the first high-altitude astronomical observatory. Interweaves an account of Floyd's life and work with a discussion of Lick, the patron of the observatory.

CHEMISTRY

268. Jones, Daniel P. "Chemical Warfare Research during World War I: A Model of Cooperative Research." *Chemistry and Modern Society: Historical Essays in Honor of Aaron J. Ihde.* Edited by John Parascandola and James C. Whorton. Washington, D.C.: American Chemical Society, 1983, pp. 165-185.

Argues that the project-research-method of organizing research by stipulating projects and then allotting them to researchers in separate laboratories was first developed in the United States during World War I for chemical warfare. Finds that after the War, the model was tried by chemists to foster cooperative research. Refuses to claim a direct link.

269. Parascandola, John. "Charles Holmes Herty and the Effort to Establish an Institute for Drug Research in Post World War I America." *Chemistry and Modern Society: Historical Essays in Honor of Aaron J. Ihde.* Edited by John Parascandola and James C. Wharton. Washington, D.C.: American Chemical Society, 1983, pp. 85-103.

Links Herty's unsuccessful labors to establish an Institute for Drug Research to free the United States from dependency on foreign countries for synthetic organic chemicals with the later establishment of the National Institutes of Health.

270. Robinson, Lisa Mae. "The Electrochemical School of Edgar Fahs Smith, 1878-1913." Ph. D. dissertation, University of Pennsylvania, 1986.

Argues that Smith's most enduring impact on American chemistry was his graduate program at the University of Pennsylvania. Characterizes his Ph.D. students as "managers of science," who learned how to build and run an institution from Smith.

271. Servos, John W. "A Disciplinary Program that Failed: Wilder D. Bancroft and the *Journal of Physical Chemistry,* 1896-1933." *Isis,* 73 (1982): 207-232.

Presents Bancroft as a representative of the view that physical chemistry was the core of chemistry. Describes his opponents as those who saw physical chemistry as the bridge between physics and chemistry. Traces Bancroft's unsuccessful effort to define the discipline.

272. Servos, John W. "History of Chemistry." *Osiris*, 2nd ser., 1
      (1985): 132-146.

      Finds the literature in the history of American chemistry
      limited compared to the history of other scientific disciplines in
      the United States. Calls for histories of a number of areas of
      chemical research, including organic, colloid, and polymer
      chemistry. Argues that chemistry is useful in illuminating general
      problems in the history of science, such as the transmission of
      scientific ideas and techniques from one national context to
      another, the growth and decay of specialties and interdisciplinary
      fields, and the role of local circumstances in shaping national
      scientific styles.

273. Servos, John W. "The Intellectual Basis of Specialization:
      Geochemistry in America, 1890-1915." *Chemistry and
      Modern Society: Historical Essays in Honor of Aaron J.
      Ihde*. Edited by John Parascandola and James C. Wharton.
      Washington, D.C.: American Chemical Society, 1983,
      pp. 1-19.

      Contends that geochemistry arose in the laboratories of the
      United States Geological Survey and the Carnegie Institution in
      response to the recognition by physical chemists and petrog-
      raphers that the techniques of the former could solve the
      problems of the latter. Argues that the willingness of the
      American scientific community to participate in cross-disciplinary
      collaborations was important for the birth of geochemistry.

274. Tarbell, Dean Stanley, and Ann Tracy Tarbell. *Essays in the
      History of Organic Chemistry in the United States, 1875-
      1955*. Nashville: Folio Publishers, 1986. x+433 pp.
      Indices.

      Stresses the growth of ideas, the interrelationship of
      organic chemistry to other fields, key personalities, and major
      institutions. Focuses on technical knowledge rather than context.
      Assumes that the reader understands the general outline of the
      history of chemistry. This is a useful reference work.

275. Tarbell, Dean Stanley, and Ann Tracy Tarbell. *Roger Adams:
      Scientist and Statesman*. Washington, D.C.: American
      Chemical Society, 1981. viii+240 pp. Appendices, Index.

      Illuminates the life of one of America's most influential
      chemists. Provides insight into the development of American
      chemical education between the World Wars. Demonstrates how

the rapid evolution of a discipline can leave an older researcher behind.

276. Thackray, Arnold, Jeffrey L. Sturchio, P. Thomas Carroll, and Robert Bud. *Chemistry in America, 1876-1976: Historical Indicators*. Dodrecht: D. Reidel, 1985. xxiii+564 pp. Appendices, Bibliography, Index.

Contains quantitative indicators which may be used by future historians. Views chemistry as an occupation, a profession, and a discipline. Presents both chemical education and the chemical industry as contexts. Over half the book consists of appendices, tables, and bibliography.

GEOLOGY

277. Blunt, John. "On the Growth of a Prescient Speculation: Glacial Geology in Nineteenth-Century America." Ph.D. dissertation, New York University, 1984.

Divides the American contributions to the growth of glacial geology and the religious debate which accompanied them into three stages delineated by the development of the theory of an Ice Age by Louis Agassiz. Argues that geologists in the United States did not accept Agassiz's theory until the emergence of a new generation in the 1850s. Believes the earlier generation rejected Agassiz's theory because these scientists had a propensity for interpreting geological data in ways which did not generally conflict with Biblical accounts.

278. Corgan, James X. "Early American Geological Surveys and Gerard Troost's Field Assistants, 1831-1836." *The Geological Sciences in the Antebellum South* (item 279), pp. 39-72.

Argues that Troost was responsible for the French influence upon pre-1836 American field geology.

279. Corgan, James X., editor. *The Geological Sciences in the Antebellum South*. University: University of Alabama Press, 1982. 195 pp. Appendix, Bibliography, Index.

Presents nine papers focusing on the work of the field geologist. Includes items 184, 278, 295.

280. Donovan, Arthur L. "New Views on the Origins of Geology."
     Northeastern Geology, 3 (1981): 2-4.

     Identifies two different approaches to the history of
     geology: either defining the discipline in terms of its subject
     matter and conceptual structure, or in terms of its social
     organization. Sees the former as emphasizing the linkages
     between geology and other sciences, and science and other
     aspects of high culture. Sees the latter as highlighting the relation
     of science to other control mechanisms.

281. Drake, Ellen T. "The Coon Butte Crater Controversy."
     *Geologists and Ideas: A History of North American
     Geology* (item 282), pp. 65-78.

     Argues that the explanations for the reluctance of the
     geological community to accept an impact explanation for this
     crater were multiple and complex.

282. Drake, Ellen T., and William M. Jordan, editors. *Geologists and
     Ideas: A History of North American Geology.* Boulder:
     Geological Society of America, 1985. x+525 pp. Index.

     Presents thirty-three papers organized around four themes:
     the history of important ideas, contributions of individuals,
     contributions of groups or institutions, and the application of
     ideas. Contains autobiographical accounts as well as historical
     discussions. Includes items 281, 288, 294, 419.

283. Eagan, William E. "'I Would Have Sworn My Life on Your
     Interpretation': James Hall, Sir William Logan and the
     'Quebec Group'." *Earth Sciences History,* 6 (1987): 47-
     60.

     Traces Logan's adoption of Hall's work in paleontology
     to interpret the rocks of Quebec and subsequent rejection when
     evidence arose challenging Hall's interpretation. Finds that
     personalities and nationalism colored the relationship between
     Hall and Logan, the director of the Geological Survey of
     Canada.

284. Eckel, Edwin B. *The Geological Society of America: Life History
     of a Learned Society.* Boulder: Geological Society of
     America, 1982. xiv+167 pp. Appendices, Bibliography.

     Focuses on the administrative history of the society.
     Stresses the influence of the 1931 bequest of Richard A. F.

Penrose, Jr.--which made the Geological Society of America one of the most wealthy learned societies in the world--on the evolution of the society.

285. Goodstein, Judith. "Waves in the Earth: Seismology Comes to Southern California." *Historical Studies in the Physical Sciences*, 14 (1984): 201-230.

Documents the rise of seismology at the California Institute of Technology between the world wars. Describes it as the result of the combination of the European tradition of using seismology as a tool in geophysical research and the American hope to use small earthquakes to predict larger ones. Credits Beno Gutenberg, Henry Wood, and Charles Richter as the movers in California seismology.

286. Greene, Mott T. "History of Geology." *Osiris*, 2nd ser., 1 (1985): 97-116.

Notes that the history of geology was taken up by historians of science later than the history of other disciplines; it is a field which has been and will probably continue to be dominated by geologists writing for other geologists. Claims that it is too early for historians to exploit archival sources because the printed sources are not yet well explored.

287. Jordan, William M., editor. "History of Geology in the Northeast: Proceedings of a Symposium." *Northeastern Geology*, 3 (1981): 1-103.

Contains fifteen papers focusing primarily on individual geologist and surveys, especially in Pennsylvania. Includes items 178, 183, 280, 290.

288. Laporte, Léo F. "Wrong for the Right Reasons: G. G. Simpson and Continental Drift." *Geologists and Ideas: A History of North American Geology* (item 282), pp. 273-285.

Argues that George Gaylord Simpson rejected continental drift because he could explain past biogeographies using a theory of evolutionary ecology which did not need drift.

289. Leviton, Alan E. and Michele L. Aldrich. "John Boardman Trask: Physician Geologist in California, 1850-1879." *Frontiers of Geological Exploration of Western North America: A Symposium*. Edited by Alan E. Leviton, Peter U. Rodda, Ellis L. Yockelson, and Michele L. Aldrich.

San Francisco: Pacific Division of the American Association for the Advancement of Science, 1982, pp. 37-69.

Argues that Trask's research prepared the way for Josiah D. Whitney's work. Includes an extensive bibliography.

290. Schneer, Cecil J. "Macrogeology and Microgeology; Theory and Empiricism in American Geology, a Discussion." *Northeastern Geology*, 3 (1981): 101-103.

Places American geology in world context. Identifies shifts between concern for macro- and micro-geological problems. Superficial.

291. Schultz, Susan. "The Debate Over Multiple Glaciation in the United States: T. C. Chamberlin and G. F. Wright, 1889-1894." *Earth Sciences History*, 2 (1983): 122-129.

Describes the clash between Chamberlin, who believed in multiple glaciation epochs, and Wright, an advocate of the theory of continuity of glaciation. Identifies a number of elements in the clash, including personal animosities, the perception of the arrogance of government science, and proprietary attitudes towards scientific knowledge.

292. Servos, John W. "To Explore the Borderland: The Foundation of the Geophysical Laboratory of the Carnegie Institution of Washington." *Historical Studies in the Physical Sciences*, 14 (1984): 147-185.

Claims that the Geophysical Laboratory has been one of the most important institutions in the field of geology. Argues that the history of the Laboratory illuminates the process which resulted in the introduction of the chemical and physical traditions in the earth sciences and the introduction of laboratory techniques into what had been a branch of natural history.

293. Shrock, Robert Rakes. *Geology at M.I.T., 1865-1965: A History of the First Hundred Years of Geology at the Massachusetts Institute of Technology.* Volume II: *Departmental Operations and Products.* Cambridge: The M.I.T. Press, 1982. xxvi + 762 + 244 pp. Appendices, Indices.

Discusses the facilities, physical plant, publications, and finances. Summarizes the scientific contributions of M.I.T. in a number of geological specialties and subdisciplines. Provides information on course content, degree recipients, thesis titles, and faculty. Emphasizes the role of women in the department. This is a continuation of Item I:463.

294. Skinner, Brian J. and Barbara L. Narendra, "Rummaging through the Attic; or, A Brief History of the Geological Sciences at Yale." *Geologists and Ideas: A History of North American Geology* (item 282), pp. 355-376.

Divides the history into seven stages, each representing a generation. Finds slow, steady growth.

295. White, George W. "Andrew Ellicott's Geological Observations in the Mississippi and Florida, 1796-1800." *The Geological Sciences in the Antebellum South* (item 279), pp. 9-25.

Derives information from travel reports. Claims that travel reports are the primary source of data for early American geology.

MATHEMATICS

296. Closs, Michael P., editor. *Native American Mathematics.* Austin: University of Texas Press, 1986. 431 pp. Bibliography.

Emphasizes Mesoamerica.

297. Cohen, Patricia Cline. *A Calculating People: The Spread of Numeracy in Early America.* Chicago: University of Chicago Press, 1982. x+271 pp. Index.

Contends that the increasing numeracy after the Revolutionary War was due both to the increasing number of people in the market economy, and the efforts to prove the success of the political revolution through statistics of growth and power. Concludes that by 1850 the prestige of quantification was soaring, with statistical knowledge viewed as objective truth. Emphasizes that people counted things that mattered to them with little regard for imprecise definitions or assumptions that subverted the count.

298. Hogan, Edward R. "Theodore Strong and Ante-Bellum American Mathematics." *Historia Mathematica,* 8 (1981): 439-455.

Presents Strong as a sophisticated, albeit unoriginal, mathematician for his time and place. Credits him with important roles both in the dissemination of advanced mathematics in the

United States and the education of significant post-bellum mathematicians.

299. Parshall, Karen Hunger. "Eliakim Hastings Moore and the Founding of a Mathematical Community in America, 1892-1902." *Annals of Science,* 41 (1984): 313-333.

Describes Moore's labors as an organizer, editor, and teacher. Credits him with being a major force in the transformation of American mathematics from minor status to world leadership.

300. Reingold, Nathan. "Refugee Mathematicians in the United States, 1933-1941: Reception and Reaction." *Annals of Science,* 38 (1981): 313-338.

Places the response of the American mathematical community to refugees mathematicians in the context of the ideology of the universality of science, Depression conditions, American nationalism and anti-Semitism, and the image of the United States as a haven for oppressed. Identifies the founding of the journal *Mathematical Reviews* and the increase of the status of applied mathematics as the two most important consequences for American mathematics.

301. Rider, Robin E. "Alarm and Opportunity: Emigration of Mathematicians and Physicists to Britain and the United States, 1933-1945." *Historical Studies in the Physical Sciences,* 15 (1984): 107-176.

Includes a master list of displaced or émigré physicists and mathematicians. Finds that they were typically male and less than 40 years old. Provides information on the various middlemen who assisted in the emigration.

302. Servos, John W. "Mathematics and the Physical Sciences in America, 1880-1930." *Isis,* 77 (1986): 611-629.

Blames the relative backwardness of American scientists in the mathematical aspects of the physical sciences to the inadequacy of American mathematical training for scientists. Argues that this inadequacy was due to the disinterest of American mathematicians in applications.

PHYSICS

303. Cornell, Thomas D. "Merle A. Tuve and His Program of Nuclear Studies at the Department of Terrestrial Magnetism: The Early Career of a Modern American Physicist." Ph.D. dissertation, Johns Hopkins University, 1986.

Traces Tuve's career through the commencement of World War II. Presents an example of physics research utilizing sophisticated technology and complex institutional structures. Sees Tuve evolving from the experimental physicist as master craftsman to the physicist as engineer/administrator.

304. Davies, Shannon M. "American Physicists Abroad: Copenhagen, 1920-1940." Ph.D. dissertation, University of Texas, 1985.

Reviews the changing relationship between Niels Bohr and the American physics community. Demonstrates that over the two decades, Americans evolved from students at the periphery of Bohr's circle to collaborators.

305. Fellows, Frederick Hugh. "J. H. Van Vleck: The Early Life and Work of a Mathematical Physicist." Ph.D. dissertation, University of Minnesota, 1985.

Discusses Van Vleck's life and work until he went to Harvard in 1934. Focuses on the publications of Van Vleck and his students on the old quantum theory and his theory of electric and magnetic susceptibilities in gases.

306. Good, Gregory A. "Geomagnetics and Scientific Institutions in 19th Century America." *Eos*, 66, no. 27 (July 2, 1985): 521, 524-526.

Tests his thesis that intellectual status is a contingency in the development of the institutional form of a science by examining American geomagnetic efforts in the mid-nineteenth century. Argues that its institutions were appropriate for an intellectually dependent field. Rejects the view of institutionalization set forth by Daniels (see Item I:239).

307. Henriksen, Paul W. "Solid State Physics Research at Purdue."
      Osiris, 2nd ser., 3 (1987): 237-260.

      Focuses on research on germanium made in response to
      World War II. Illustrates how basic and applied research
      followed different paths. Argues that Purdue's research program
      was in reaction to funding opportunities.

308. Hoch, Paul K. "The Reception of Central European Refugee
      Physicists of the 1930s: U.S.S.R., U.K., U.S.A." *Annals
      of Science*, 40 (1983): 217-246.

      Rejects earlier claims that the integration of these
      physicists was straight-forward. Concludes that the United States
      was able to absorb refugees when the other nations could not
      because of an expanding educational system.

309. Hoddeson, Lillian. "Establishing KEK in Japan and Fermilab in
      the US: Internationalism, Nationalism, and High Energy
      Accelerators." *Social Studies of Science*, 13 (1983): 1-
      48.

      Compares and contrasts the Fermi National Accelerator
      Laboratory (Fermilab) and the Japanese high energy physics
      laboratory, Kō Enerugii Butsurigaku Kenkyusho (KEK).
      Attributes many of the parallels of their structures to the
      internationalism of high energy physics. Attributes divergences
      to national differences. Identifies the relative conservatism of
      Japan's physics community and the greater complexity of the
      Japanese decision-making process as among the most significant
      national differences.

310. Hoddeson, Lillian. "The First Large-scale Application of
      Superconductivity: The Fermilab Energy Doubler, 1972-
      1983." *Historical Studies in the Physical and Biological
      Sciences*, 18 (1987): 25-54.

      Characterizes the Fermilab approach as rapid production
      and testing of only partially understood prototypes. Credits
      Fermilab's success to the strength of the administrative commit-
      ment of the laboratory to the project, a willingness to invest in
      the project and make it a priority, the combination of cut-and-
      try engineering and scientific research, and flexibility. Sees the
      success of such innovations as dependent both upon scientific
      talent and managerial skill.

311. Holton, Gerald. "The Formation of the American Physics Community in the 1920s and the Coming of Albert Einstein." *Minerva,* 19 (1981): 569-581.

    Argues Einstein chose to immigrate to the United States because of the quality of the indigenous scientific community. Points to the extent of private philanthropy and the hospitality to new ideas as positive aspects of American science in the 1920s. Concludes that even before the influx of European physicists in the 1930s, American physics was sound and productive.

312. Kargon, Robert H. "Henry Rowland and the Physics Discipline in America." *Vistas in Astronomy,* 29 (1986): 131-136.

    Describes Rowland as the "father of the physics discipline in America." Highlights the role of the laboratory and mental discipline in Rowland's concept of graduate training.

313. Kargon, Robert H. *The Rise of Robert Millikan: Portrait of a Life in American Science.* Ithaca: Cornell University Press, 1982. 203 pp. Bibliographical Note, Index.

    Portrays Millikan as a conservative scientist caught up in revolutionary times. Uses Millikan's life to illustrate and illuminate various themes in the history of American science. Describes Millikan's life as a microcosm of the new roles taken up by American scientists in the twentieth century. Excels in its descriptions of Millikan as an institution builder and as institutional history, but this is not a definitive biography of Millikan.

314. Kupperman, Karen Ordahl. "The Puzzle of the American Climate in the Early Colonial Period." *American Historical Review,* 87 (1982): 1262-1289.

    Traces the evolution of European ideas about climate in the wake of European experiences in North America. Argues that the discovery that climate was not consistent with latitude threatened the colonization effort.

315. Moyer, Albert E. *American Physics in Transition: A History of Conceptual Change in the Late Nineteenth Century.* Los Angeles: Tomash, 1983. xx + 218 pp. Bibliography, Index.

    Addresses the conceptual disturbances that occurred in American physics from 1870 to 1905 as it shifted from unity to multiplicity, away from the atomo-mechanical theory. Focuses on intellectual history. This is a revised version of Item I:514.

316. Moyer, Albert E. "History of Physics." *Osiris*, 2nd ser., 1 (1985): 163-182.

    Labels most history of American physics as "Manhattan-style history," which he identifies as institutional history with concern for physicist-politicians and their research problems. Suggests as an alternative the history of solid-state physics, which would demonstrate the links between physics and industry, and the survival of modest, autonomous research during a period of preoccupation with big science.

317. Rosenberg, Robert. "American Physics and the Origins of Electrical Engineering." *Physics Today*, 36 (October 1983): 48-54.

    Discusses the growth of electrical engineering out of physics at Cornell and the Massachusetts Institute of Technology. Observes that electricity gave physics research a utilitarian justification.

318. Schweber, S. S. "The Empiricist Temper Regnant: Theoretical Physics in the United States, 1920-1950." *Historical Studies in the Physical and Biological Sciences*, 17 (1986): 55-98.

    Describes the transformation of study of theoretical physics (quantum theory) in the United States. Credits the spectacular metamorphosis in the 1920s to a policy created by the leading American experimentalists and funded by the foundations. Characterizes American theoretical physics as empirical, pragmatic, and instrumentalist. Argues that the integration of theoreticians and experimentalists in American universities shaped American theoretical physics.

319. Schweber, S. S. "Shelter Island, Pocono, and Oldstone: The Emergence of American Quantum Electrodynamics after World War II." *Osiris*, 2nd ser., 2 (1986): 265-302.

    Advocates that these conferences set the stage for the rapid and substantial advances in post-war quantum mechanics by bringing together the experimental and theoretical communities.

320. Sherman, Michael. "Oppenheimer: What a Trouble-Maker!"
     The Public Historian, 4 (1982): 97-117.

     Worries that the significance of Oppenheimer's career to
     understanding crucial issues of the twentieth century may blind
     historians to the significance of his research. Reviews four books
     which focus on Oppenheimer the man and scientist, and on life
     at Los Alamos.

321. Stuewer, Roger H. "Nuclear Physicists in a New World: The
     Émigrés of the 1930s in America." *Berichte zur
     Wissenschaften,* 7 (1984): 23-40.

     Overview. Emphasizes the importance of the publication
     of three articles in the *Reviews of Modern Physics* by Hans Bethe
     and associates which reviewed the state of theoretical knowledge
     in nuclear physics.

322. Walter, Maila L. Koljonen. "Science and Cultural Crisis: An
     Intellectual Biography of Percy Williams Bridgman."
     Ph. D. dissertation, Harvard University, 1985.

     Places the intellectual life of this Harvard physicist within
     the framework of the philosophical crisis created by relativity and
     quantum mechanics and the ideological crisis resulting from
     economic depression and war. Sees operational analysis as
     Bridgman's effort to cope with these crises.

     Published as *Science and Cultural Crisis: An Intellectual
     Biography of Percy Williams Bridgman (1882-1961).* Stanford:
     Stanford University Press, 1990. xii+362pp. Bibliography,
     Index.

CHAPTER V: THE BIOLOGICAL SCIENCES

BOTANY

323. Berkeley, Edmund, and Dorothy Smith Berkeley. *The Life and Travels of John Bartram: From Lake Ontario to the River St. John.* Tallahassee: University Presses of Florida, 1982. xvi+376 pp. Appendices, Bibliography, Index.

Provides a detailed, dependable biography of the colonial botanist. Highlights his interest beyond botany. Characterizes Bartram as a keen observer.

324. Haygood, Tamara Miner. *Henry William Ravenel, 1814-1887: South Carolina Scientist in the Civil War Era.* Tuscaloosa: The University of Alabama Press, 1987. xii+204 pp. Bibliography, Index.

Presents Ravenel--slaveholder, planter, aristocrat, and botanist--as the exemplar of the antebellum Southern scientist. Traces the transformation of Ravenel from avocational scientist to professional plant collector. Rejects arguments that the South was intrinsically unsuitable for science. Argues that antebellum Southern science was nearly on a par with that of the rest of the nation. Suggests that postbellum Southern science was further behind because of the adverse economic impact of the Civil War.

325. Keeney, Elizabeth Barnaby. "The Botanizers: Amateur Scientists in Nineteenth-Century America." Ph.D. dissertation, University of Wisconsin, Madison, 1985.

Argues that differences in motivation distinguished amateur and professional botanists in the nineteenth century: professionals sought to advance or diffuse knowledge, while

amateurs botanize for personal enrichment and pleasure. Sees
the relationship between professional and amateur botanists as
mutually beneficial as long as professional research was generally
taxonomic. Finds that the communities' interests diverged in the
late nineteenth century, in the wake of the professional embrace-
ment of the new botany, with its dependence upon experimentation, with amateur botanists then turning to nature study.

326. Slack, Nancy G. "Nineteenth-Century American Women
      Botanists: Wives, Widows, and Work." *Uneasy Careers
      and Intimate Lives: Women in Science, 1789-1979.* Edited
      by Pnina G. Abir-Am and Dorinda Outram. New
      Brunswick: Rutgers University Press, 1987, pp. 77-103.

      Analyzes the relationship between marriage and botanical
work for twenty-one women. Identifies four patterns: botanize
before marriage, but not afterwards; never marry and botanize
throughout one's life; attempt to combine botanizing with a
marriage and children; and botanize after widowhood. Finds that
the single women were among the most productive, while
combining science and the raising of children was very difficult.

327. Slack, Nancy G. "Charles Horton Peck, Bryologist, and the
      Legitimation of Botany in New York State." *Memoirs of
      the New York Botanical Garden,* 1987, 45: 28-45.

      Uses Peck's correspondence to throw light on state
support of scientific research, the lack of professionalization of
bryology in the nineteenth century, and Peck's contributions to
the field.

328. Tobey, Ronald C. *Saving the Prairies: The Life Cycle of the
      Founding School of American Plant Ecology, 1895-1915.*
      Berkeley: University of California Press, 1981. x + 315
      pp. Appendix, Bibliography, Index.

      Argues that the content of the scientific knowledge of the
grassland ecologists of the Midwest who shared the experience
of the prairies was related to the social structure of the community and their historical role. Correlates their changing views
towards vegetative change to changing practical needs. Focuses
on the work of Frederic Clement, his intellectual roots, and his
influence. Contends that the grasslands ecology community
withstood intellectual attack from without but declined from
within.

329. Volberg, Rachel Ann. "Constraints and Commitments in the Development of American Botany, 1880-1920." Ph.D. dissertation, University of California, San Francisco, 1983.

Approaches the professionalization of botany from the perspective of the sociology of science. Argues that during professionalization, problems not responsive to experimental methods were abandoned.

EUGENICS AND GENETICS

330. Allen, Garland E. "The Eugenics Record Office at Cold Spring Harbor, 1910-1940: An Essay in Institutional History." *Osiris*, 2nd ser. 2 (1986): 225-264.

Contends that the establishment of the Eugenics Record Office was pivotal to the development of eugenics in the United States; it acted as a clearing house and repository for data. Sees the financial support for the Eugenics Record Office resulting from a general climate supportive of economic and social control.

331. Allen, Garland E. "The Misuse of Biological Hierarchies: The American Eugenics Movement, 1900-1940." *History and Philosophy of the Life Sciences*, 5 (1983): 105-128.

Contends that eugenicists attempted to create a rank-ordered hierarchy in the guise of an aggregational hierarchy.

332. Allen, Garland E. "Thomas Hunt Morgan: Materialism and Experimentalism in the Development of Modern Genetics." *Social Research*, 51 (1984): 709-738.

Identifies the revolution in biology during the period 1880-1930 as a shift from idealism, description, and speculation, to materialism and experimentalism. Credits Morgan with bringing together cytology and breeding experiments; also credits him with institutionalizing mechanistic biology at the California Institute of Technology.

333. Carlson, Elof Axel. *Genes, Radiation, and Society: The Life and Work of H. J. Muller.* Ithaca: Cornell University Press, 1981. xiv+457 pp. Index.

Intertwines an account of Muller's life with the history of the development of the Mendelian-chromosome theory of heredity. Discusses Muller's concern for the use of biology to

improve human life though eugenics, and his warnings about the
dangers of x-ray radiation and fall-out. Utilizes interviews with
the subject and his widow, as well as other oral and archival
sources; written with Muller's cooperation.

334. Harwood, Jonathan. "National Styles in Science: Genetics in
      Germany and the United States between the World Wars."
      Isis, 78 (1987): 390-414.

Finds that American scientists defined the scope of a
discipline more restrictively than the Germans. Credits this to the
encouragement of specialization and narrow discipline conception
by expanding American universities and research institutions.
Contrasts this to the stagnant German universities, whose
organization and lack of growth hindered the institutionalization
of new disciplines. Admits that his model is time specific and
more likely to be correct for weakly institutionalized disciplines.

335. Keller, Evelyn Fox. *A Feeling for the Organism: The Life and
      Work of Barbara McClintock.* New York: W. H. Freeman
      and Company, 1983. xix+235 pp. Glossary, Index.

Argues that McClintock's theory of transposition remained
at the periphery of genetics for so long because both her
vocabulary and approach to science was unintelligible to many
of her colleagues. Raises the question whether McClintock's
personal and professional eccentricities would have been more
readily accepted if she had been male. Relies heavily on
interviews with the subject.

336. Kimmelman, Barbara A. "The American Breeders' Association:
      Genetics and Eugenics in an Agricultural Context, 1903-
      13." *Social Studies of Science,* 13 (1983): 163-204.

Studies the first national, membership-based American
institution to promote genetic and eugenic research. Finds
relationships between agriculture, genetics, and eugenics. Argues
that the links between the ABA constituency and the Country
Life movement--a nativist movement for reforming rural life--
were decisive in the speedy institutional development of
American eugenics.

337. Kimmelman, Barbara A. "A Progressive Era Discipline: Genetics
      at American Agricultural Colleges and Experiment
      Stations, 1900-1920." Ph. D. dissertation, University of
      Pennsylvania, 1987.

Considers the development of autonomous departments dedicated to basic research in genetics at three agricultural college/experiment station complexes.

338. Mendelsohn, Everett. "'Frankenstein at Harvard': The Public Politics of Recombinant DNA Research." *Transformation and Tradition in the Sciences: Essays in Honor of I. Bernard Cohen* (item 030), pp. 317-335.

Discusses the 1977 Cambridge City Council restrictions on recombinant DNA research. Evaluates the event in terms of the social assessment and social regulation of scientific activity by those outside the scientific community.

339. Ogilvie, Marilyn Bailey, and Clifford J. Choquette. "Nettie Maria Stevens (1861-1912): Her Life and Contributions to Cytogenetics." *Proceedings of the American Philosophical Society,* 125 (1981): 292-311.

Presents her as a representative of the nineteenth-century woman able to take advantage of the new opportunities for training in science. Emphasizes her theory of chromosomal sex determination. Concludes she did not receive the recognition due her.

NATURAL HISTORY

340. Ainley, Marianne Gosztonyi. "Field Work and Family: North American Women Ornithologists, 1900-1950." *Uneasy Careers and Intimate Lives: Women in Science, 1789-1979.* Edited by Pnina G. Abir-Am and Dorinda Outram. New Brunswick: Rutgers University Press, 1987, pp. 60-76.

Provides examples of three women: one unmarried, one married with a subordinate status in the marriage, and one married, but with a shared relationship. Finds that women did have a problem balancing the demands of family with the necessity of field research.

341. Anderson, H. Allen. *The Chief: Ernest Thompson Seton and the Changing West.* College Station: Texas A&M University Press, 1986. xii+363 pp. Bibliography, Index.

Views Seton's dog stories as typical Western stories with animals substituting for the traditional lone mountain man or gunman. Blames the Nature Faker controversy on Seton's attempts to inject morality into his stories. Argues that the controversy drove Seton to improve his reputation as a scientist. Defends Seton's scientific accuracy.

342. Benson, Maxine. *Martha Maxwell: Rocky Mountain Naturalist.* Lincoln: University of Nebraska Press, 1986. xix+335 pp. Appendix, Bibliography, Index.

Documents the story of a nineteenth-century female naturalist-taxidermist who gained fame exhibiting at the Centennial Exposition. Attributes her accomplishments in taxidermy and exhibition to trial and error.

343. Coan, Eugene. *James Graham Cooper: Pioneer Western Naturalist.* Moscow, Idaho: The University Press of Idaho, 1981. 255 pp. Appendices, Bibliography, Index.

Presents the life and work of a physician/naturalist who worked on the Pacific Railroad Survey as one of S. F. Baird's collectors. Demonstrates the importance of federal or state assistance for the conduct of science in nineteenth-century America.

344. Cutright, Paul Russell, and Michael J. Brodhead. *Elliott Coues: Naturalist and Frontier Historian.* Urbana: University of Illinois Press, 1981. xv+509 pp. Appendices, Bibliography, Index.

Relates the life of an ornithologist and pioneer documentary editor. Demonstrates how the career of a surgeon in the United States Army was conducive to scientific research. Provides considerable evidence of the instability of Coues's personality.

345. Deiss, William A. "The Making of a Naturalist: Spencer F. Baird, the Early Years." *From Linnaeus to Darwin: Commentaries on the History of Biology and Geology.* Edited by Alwyne Wheeler and James H. Price. London: Society for the History of Natural History, 1985, pp. 141-148.

Focuses on Baird's life from 1840 through 1846. Concludes that the preconditions for success as a naturalist included access to books, specimens, and other naturalists; financial support; leisure to pursue research; and the willingness of society to accept this activity as worthwhile.

346. Deiss, William A. "Spencer F. Baird and His Collectors."
Journal of Society for the Bibliography of Natural History,
9 (1980): 635-645.

Discusses Baird's efforts to organize, equip, and support
field collectors while serving as Assistant Secretary of the
Smithsonian Institution.

347. Fitzpatrick, Thomas Jefferson. *Rafinesque: A Sketch of His Life
with Bibliography.* Revised and Enlarged by Charles
Boewe. Weston, Mass.: M&S Press, 1982. vi+360 pp.
Bibliography.

Corrects both the biographical sketch and bibliography.
Includes guides to manuscript collections and secondary
literature.

348. Kohlstedt, Sally Gregory. "Collectors, Cabinets and Summer
Camp: Natural History in the Public Life of Nineteenth-
Century Worcester." *Museum Studies Journal,* 2 (1985):
10-23.

Presents Worcester as an example of the public museum
movement in smaller cities. Uses the case study to illuminate the
effectiveness of both public inducements and private yearnings.
Discusses the important role of articulate and successful scientists
in persuading community leaders and philanthropists to support
natural history museums.

349. Kohlstedt, Sally Gregory. "Henry A. Ward: The Merchant
Naturalist and American Museum Development." *Journal
of the Society for the Bibliography of Natural History,* 9
(1980): 647-661.

Offers Ward as an example of a facilitator of scientific
investigation and popularization.

350. Kohlstedt, Sally Gregory. "International Exchange and National
Style: A View of Natural History Museums in the United
States, 1850-1900." *Scientific Colonialism: A Cross-
Cultural Comparison.* Edited by Nathan Reingold and
Marc Rothenberg. Washington, D.C.: Smithsonian
Institution Press, 1987, pp. 167-190.

Concludes that American museum staff attempted to meet
European standards while remaining responsive to local sponsors.

351. Kohlstedt, Sally Gregory. "Natural History at Dickinson and Other Colleges in the Nineteenth Century." *John and Mary's Journal,* 1985 (10): 27-48.

     Divides the history of natural history on American college campuses into three periods: intermittent activity, through about 1822; increased informal and some formal activity, to about 1870; and the establishment of systematic collections and a scientific curriculum, along with the erection of buildings, 1870-1900.

352. Meyers, Amy R. Weinstein. "Sketches from the Wilderness: Changing Conceptions of Nature in American Natural History Illustration: 1680-1880." Ph.D. dissertation, Yale University, 1985.

     Focuses on five naturalists-artists: John Banister, Mark Catesby, William Bartram, Titian Ramsey Peale, and Timothy H. O'Sullivan. Describes the images they produced as not only reflecting prevailing attitudes towards the natural world, but also serving as catalysts for change. Argues that these men made their greatest impact on the course of science in the field of environmental thought.

353. Miller, David Stuart. "An Unfinished Pilgrimage: Edwin Way Teale and American Nature Writing." Ph.D. dissertation, University Minnesota, 1982.

     Presents Teale as a representative American nature writer. Describes his writings as attempts to have his readers accept that the natural world is characterized by struggle and death on one hand, and beauty and interest on the other.

354. Porter, Charlotte M. *The Eagle's Nest: Natural History and American Ideas, 1812-1842.* University: University of Alabama Press, 1986. xii+251 pp. Bibliography, Index.

     Assesses the role of a group of naturalists prominent in the Academy of Natural Sciences of Philadelphia and New Harmony, Indiana, in the development of American natural history. Offers a sympathetic view of the efforts of field naturalists. Fails to link the publications of these individuals to larger American ideas.

355. Porter, Charlotte M. "The Lifework of Titian Ramsey Peale." *Proceedings of the American Philosophical Society,* 129 (1985): 300-312.

Presents his career as an exemplification of the associations and connections within the naturalist community in Philadelphia during the first half of the nineteenth century. Describes him as an artist with little literary skill. Concludes that he had many opportunities, but failed to fully exploit them.

356. Sorensen, Willis Conner. "Brethren of the Net: American Entomology, 1840-1880." Ph.D. dissertation, University of California, Davis, 1984.

Credits the rise of American entomology to world leadership to the establishment of institutions and collections, state and federal support for research into insect control, and the framework provided by evolutionary theory after 1859.

357. Spencer, Larry T., et al. "Naturalists and Natural History Institutions of the American West." *American Zoologist,* 26 (1986): 295-384.

Contains eleven short essays covering the period from Lewis and Clark through the early twentieth century. Argues that the amateur tradition in natural history was the root of American professional biology. Sees the only major difference in the development of biology in the American East and the American West as chronological.

358. Tobin, Mary Frances. "Nature Writers as Dissenting Moderns: Modernization and the Development of American Beliefs about Nature." Ph.D. dissertation, University of Maryland, 1981.

Argues that twentieth-century nature writers have not rejected modern society and its values, despite appearances to the contrary.

PALEONTOLOGY

359. Rainger, Ronald. "The Continuation of the Morphological Tradition: American Paleontology, 1880-1910." *Journal of the History of Biology,* 14 (1981): 129-158.

Finds evidence of interest in morphology in the work of Edward Drinker Cope, O. C. Marsh, and Alpheus Hyatt. Argues that the generation of paleontologists following these three men continued that interest. Rejects the conclusion of Garland Allen that there was a "revolt from morphology" during this period.

360. Rainger, Ronald. "Just Before Simpson: William Diller Mat-
     thew's Understanding of Evolution." *Proceedings of the
     American Philosophical Society,* 130 (1986): 453-476.

     Argues that Matthew's training in geology distinguished
     him from his contemporaries in vertebrate paleontology, who had
     a background in biology. Believes his relative disregard of
     internal biological mechanisms and operations allowed him to
     ignore the neo-Lamarckianism of his contemporaries in favor of
     Darwinian evolution. Sees Matthew as the exception to the
     generalization that American paleontologists were unaware of or
     hostile to developments in other areas of biology.

PHYSIOLOGY

361. Appel, Toby A. "Biological and Medical Societies and the
     Founding of the American Physiological Society."
     Physiology in the American Context, 1850-1940 (item
     370), pp. 155-176.

     Locates the roots of the American Physiological Society,
     founded in 1887, partly in the conflict between the American
     Medical Association and more exclusive medical specialty
     societies, and partly in the American Society of Naturalists, the
     first of a series of biological specialty societies. Notes that the
     American Physiological Society was the first American national
     society to require publication of original research as a prere-
     quisite of membership. Sees the establishment of this prerequisite
     as an effort by elite research physicians to keep out clinical
     physiologists.

362. Benison, Saul, A. Clifford Barger, and Elin L. Wolfe. *Walter B.
     Cannon: The Life and Times of a Young Scientist.*
     Cambridge: Harvard University Press, 1987. xiv+520
     pp. Bibliography, Index.

     Considers both the personal life and the scientific work
     of this distinguished physiologist and reformer of medical
     education through his mid-40s (1917). Provides context through
     the history of physiology and medical education at Harvard
     University. Discusses his defense of medical research against
     repeated attacks by the antivivisectionist movement in America
     and his efforts to develop guidelines for human experimentation.
     Emphasizes the significance of his marriage to his life.

363. Borell, Merriley. "Instruments and an Independent Physiology: The Harvard Physiological Laboratory, 1871-1906." *Physiology in the American Context, 1850-1940* (item 370), pp. 293-321.

Highlights the work of William Townsend Porter in laboratory instruction and Porter's establishment of the Harvard Apparatus Company to mass produce laboratory equipment.

364. Brobeck, John R., Orr E. Reynolds, and Toby A. Appel, editors. *History of the American Physiological Society: The First Century: 1887-1987*. Bethesda, Maryland: American Physiological Society, 1987. viii+533 pp. Appendices, Index.

Includes both historical analysis and anecdotal remembrances. Provides a history of the society, biographical sketches of the officers, and discussions of special themes.

365. Clarke, Adele E. "Research Materials and Reproductive Science in the United States, 1910-1940." *Physiology in the American Context, 1850-1940* (item 370), pp. 323-350.

Discusses the demand for a new infrastructure of organizations to supply needed material that resulted from the expansion of physiological research. Focuses on mammalian reproductive physiology. Observes that the demand included both live animals and fresh animal parts. Shows that one result was the development of colonies of research animals.

366. Cross, Stephen J., and William R. Albury. "Walter B. Cannon, L. J. Henderson and the Organic Analogy," *Osiris*, 2nd ser., 3 (1987): 165-192.

Examines the models of social stability developed by these two physiologists in response to the Great Depression. Contrasts Cannon's support for a regulated economy with Henderson's criticism of social engineering and the New Deal. Rejects the claim that there was a decline in the use of biological modes of explanation in the human sciences around World War I.

367. Frank, Robert G., Jr. "American Physiologists in German Laboratories, 1865-1914." *Physiology in the American Context, 1850-1940* (item 370), pp. 11-46.

Finds that the leaders in American physiology pre-1920 were disproportionately those with European training. Concludes that they returned with an appreciation of the German sense of preciseness in method.

368. Fry, W. Bruce. *The Development of American Physiology: Scientific Medicine in the Nineteenth Century.* Baltimore: Johns Hopkins University Press, 1987. xi+308 pp. Appendices, Bibliography, Index.

Studies the transformation of physiology during the second half of the nineteenth century, its increasing significance in medical schools, and its role in the rise of the research ethic in American medicine. Focuses on the careers of four pioneers: John Call Dalton, Jr., S. Weir Mitchell, Henry P. Bowditch, and H. Newell Martin. Sees physiology as a link between larger reforms in American education and efforts to make medical education more scientific.

369. Geison, Gerald L. "International Relations and Domestic Elites in American Physiology, 1900-1940." *Physiology in the American Context, 1850-1940* (item 370), pp. 115-154.

Credits the rise of American physiology in part to the impact of World War I on the British and German physiology communities, and in part to the increase in the quantity and quality of American physiological laboratories post-World War I. Finds correlations between research productivity and disciplinary eminence. Supports Derek J. de Solla Price's conclusion that research productivity and achievement are unevenly distributed.

370. Geison, Gerald, L., editor. *Physiology in the American Context, 1850-1940.* Baltimore: American Physiological Society, 1987. viii+403 pp. Indices.

Focuses on institutional history, the relationship of physiology to other fields, and laboratory materials, techniques, and instruments. Concentrates on the appearance, development, and rise of physiology as an independent discipline in the United States. Contains fifteen papers. Includes items 361, 363, 365, 367, 369, 371, 372, 373, 375.

371. Gillespie, Richard P. "Industrial Fatigue and the Discipline of Physiology." *Physiology in the American Context, 1850-1940* (item 370), pp. 237-262.

Characterizes industrial fatigue as the leading area of applied physiology during this period. Describes the competition with psychologists and physicians for the role of expert. Observes that management was unhappy with the physiologists' emphasis on altering the work environment to suit the worker. Notes that difficulties arose in applying physiological principles to work.

372. Laszlo, Alejandra C. "Physiology of the Future: Institutional Styles at Columbia and Harvard." *Physiology in the American Context, 1850-1940* (item 370), pp. 67-96.

Contrasts Walter B. Cannon's success at Harvard with Frederic S. Lee's conflicts with the clinicians at Columbia. Credits Cannon's accomplishments to his argument that physiology could be used to train physicians to think scientifically and was integral to the medical curriculum. Compares this to Lee's vision of physiology as a biological science.

373. Maienschein, Jane. "Physiology, Biology, and the Advent of Physiological Morphology." *Physiology in the American Context, 1850-1940* (item 370), pp. 177-193.

Discusses the efforts of Charles O. Whitman to institutionalize physiological morphology at the Marine Biological Laboratory and the University of Chicago. Characterizes the University of Chicago as the exception to the marginalization of physiological morphology at the end of the nineteenth century.

374. Numbers, Ronald L. and William J. Orr, Jr. "William Beaumont's Reception at Home and Abroad." *Isis,* 72 (1981): 590-612.

Observes that Beaumont's work was acclaimed in the United States for its insights into practical dietetics, while Europeans focused on the physiological theory and experimental techniques. Argues that the American reception was due to the fact that almost all American physiologists were practicing physicians, not experimental scientists, and were more interested in application than the impact on theory.

375. Pauly, Philip J. "General Physiology and the Discipline of Physiology, 1890-1935." *Physiology in the American Context, 1850-1940* (item 370), pp. 195-207.

Characterizes American general physiologists as indifferent entrepreneurs with differing outlooks, united only in their hostility to medical physiology and a concern for understanding the problem of life.

376. Pitcock, Cynthia DeHaven. "The Career of William Beaumont, 1785-1853: Science and the Self-made Man in America." Ph.D. dissertation, Memphis State University, 1985.

Provides an overview of his life and research. Emphasizes his limited training and resources.

OTHER DISCIPLINES

377. Allen, Garland. "Morphology and Twentieth-Century Biology: A Response." *Journal of the History of Biology,* 14 (1981): 159-176.

Responds to items 359, 379, 389. Agrees that the term "morphology" is problematic. Admits taking the words of biologists too literally. Recognizes that the revolt against morphology was against a methodology, not a subject matter. Contends that it was a revolution, not an evolution.

378. Benson, Keith R. "American Morphology in the Late Nineteenth Century: The Biology Department at Johns Hopkins University." *Journal of the History of Biology,* 18 (1985): 163-205.

Finds that American biology developed its singular characteristics, constructed its major institutions, and established its primary educational programs at the time morphology was the dominant subject in American biology. Credits William Keith Brooks at Johns Hopkins with having a profound influence on American biology through his students.

379. Benson, Keith R. "Problems of Individual Development: Descriptive Embryological Morphology at the Turn of the Century." *Journal of the History of Biology,* 14 (1981): 115-128.

Concludes that the morphology practiced by William K. Brooks evolved from a speculative, unfocused science to a more precise critique of the germ-layer doctrine. Suggests that morphology and physiology were complementary sciences.

380. Churchill, Frederick B. "In Search of the New Biology: An Epilogue." *Journal of the History of Biology,* 14 (1981): 177-191.

Summarizes items 359, 377, 379, 389. Criticizes the paucity of discussion of the American cultural and social context that nurtured the changes in the scientific programs.

381. Edgar, Robert K. "The Origin of Diatom Biology in America." *Occasional Papers of the Farlow Herbarium*, 16 (1981): 43-58.

Focuses on the activities of Jacob Whitman. Identifies three major scientific issues: whether diatoms were animals or plants, their classification, and fossil diatoms and uniformitarian geology.

382. Greenfield, Theodore J. "Variation, Heredity, and Scientific Explanation in the Evolutionary Theories of Four American Neo-Lamarckians, 1867-1897." Ph.D. dissertation, University of Wisconsin-Madison, 1986.

Focuses on Alpheus Packard, Alpheus Hyatt, Edward Cope, and John Ryder. Argues that these men were more interested in explaining variation than in heredity. Stresses that in response to Weismann's anti-Lamarckian theory of heredity, their dominant concern was defending external causal explanations of variation, not the inheritance of acquired characteristics.

383. Kay, Lily E. "Conceptual Models and Analytical Tools: The Biology of Physicist Max Delbrück." *Journal of the History of Biology*, 18 (1985): 207-246.

Finds that Delbrück's apparent rapid rise was in fact gradual. Argues that the rise was sustained by a research commitment that biology was a branch of physics. Credits his successful career in part to his association with networks of scientists and his personal qualities.

384. Kay, Lily E. "Cooperative Individualism and the Growth of Molecular Biology at the California Institute of Technology, 1928-1953." Ph.D. dissertation, 1987.

Deals with the birth and development of molecular biology--a physico-chemical biology characterized by cooperative research utilizing expensive and sophisticated technology. Credits the Rockefeller Foundation with providing the financial support which shaped the path of molecular biology.

385. Kay, Lily E. "W. M. Stanley's Crystallization of the Tobacco
      Mosaic Virus, 1930-1940." *Isis*, 77 (1986): 450-472.

      Argues that Stanley's demonstration was well received
      despite flaws, technical errors, and misconceptions, because it
      fit with the views and expectations of a dominant group of
      scientists in protein chemistry and molecular biology during the
      1930s and 1940s. Concludes that the status of the Rockefeller
      Institute gave Stanley's work power and authority.

386. McClintock, James I. "Joseph Wood Krutch and William Morton
      Wheeler--Metabiologists." *Journal of American Culture*,
      10, 4 (Winter 1987): 37-43.

      Contends that Wheeler's work provided the scientific
      authority for Krutch's philosophy. Describes Krutch's nature
      essays as the medium for expressing his philosophy that humans
      were linked to the rest of creation.

387. Maienschein, Jane. "Experimental Biology in Transition:
      Harrison's Embryology, 1895-1910." *Studies in the
      History of Biology*, 6 (1983): 107-127.

      Presents Ross Harrison's experimental embryology as an
      exemplar of the development of experimental biology generally
      at the turn of the century. Argues that experimentalism comes in
      many forms. Contends that experimental science evolved during
      this period. Offers an alternative to the thesis of Gar Allen and
      William Coleman that biology during this period was a revolt
      from morphology to experimentation.

388. Maienschein, Jane. "History of Biology." *Osiris*, 2nd ser.,
      (1985): 147-162.

      Finds interest in American biology by historians of science
      to be a recent development. Approves the great diversity in
      approaches.

389. Maienschein, Jane. "Shifting Assumptions in American Biology:
      Embryology, 1890-1910." *Journal of the History of
      Biology*, 14 (1981): 89-113.

      Explores the range of methodologies, ontologies, problems
      addressed, and types of results sought by turn-of-the-century
      American biologists. Finds that they fall within a range.
      Examines in some detail the work of Edmund Beecher Wilson,
      Edwin Grant Conklin, Thomas Hunt Morgan, and Ross Granville

Harrison. Concludes that calling the changes either revolutionary or evolutionary distorts the picture. Maintains that shifts in individual assumptions were gradual, while shifts of research programs seemed rapid.

390. Maienschein, Jane, Ronald Rainger, and Keith R. Benson. "Introduction: Were Americans Morphologists in Revolt?" *Journal of the History of Biology,* 14 (1981): 83-87.

Introduces items 359, 377, 379, 389. Challenges Gar Allen's thesis that American biology at the turn of the twentieth century was a shift to experimentalism from morphology. Rejects Allen's subsequent claim of a naturalist-experimentalist dichotomy. Stresses continuities and gradual change.

391. Manning, Kenneth R. *Black Apollo of Science: The Life of Ernest Everett Just.* New York: Oxford University Press, 1983. 397 pp. Index.

Illustrates the difficulties facing minorities in pursuing a scientific career in the United States during the first third of the twentieth century through a well-researched and balanced biography of one of the leading experimental biologists. Demonstrates the role foundations had in the progress of scientific careers during the period between the World Wars. Analyzes both the racism of the white scientific community and Just's personality flaws in assessing his career.

392. Oppenheimer, Jane M. "Louis Agassiz as a Early Embryologist." *Science and Society in Early America: Essays in Honor of Whitfield J. Bell, Jr.* (item 25), pp. 393-414.

Credits Agassiz with being the first embryologist in the United States.

393. Pauly, Philip J. "The Appearance of Academic Biology in Late Nineteenth-Century America." *Journal of the History of Biology,* 17 (1984): 369-397.

Argues that the introduction of biology into American universities was part of the effort by reformers to raise the standard of medical education. Credits the creation of biology as a core discipline to C. O. Whitman at the University of Chicago. Concludes that where medical education did not predominate, biology thrived; where the medical school was strong, biology was ephemeral.

394. Pauly, Philip J. *Controlling Life: Jacques Loeb and the Engineering Ideal in Biology.* New York: Oxford University Press, 1987. vii+252 pp. Glossary, Index.

Rejects the contention that Loeb's primary concern was the demonstration of mechanistic materialism. Sees Loeb's program as an effort to control biological processes. Describes Loeb as a major champion of the "engineering standpoint" in biology, a view which subordinated analysis to experiment and action, and repudiated concepts which did not increase scientific control over biological activity. Concludes that Loeb was significant as a model of the reductionist scientist. Traces his importance for a group of eminent twentieth-century life scientists, including Paul De Kruif, John B. Watson, and B. F. Skinner. Argues that these Loebians shared a social marginality relative to the dominant groups in American academic science before World War II, a dedication to science, enthusiasm for experimentation, and the belief that biology could be devised as an engineering science.

395. Pauly, Philip J. "The Loeb-Jennings Debate and the Science of Animal Behavior." *Journal of the History of the Behavioral Sciences,* 17 (1981): 504-515.

Describes the competing views of Jacques Loeb and Herbert Spencer Jennings on the nature of tropisms in invertebrates during the first decade of the twentieth century. Contrasts Loeb's efforts to control behavior with Jennings's search for explanations which would account for evolutionary adaptation.

396. Pearce, Ruth P. "American Biology in a Mechanistic World: The Search for a Progressive Synthesis." Ph.D. dissertation, University of Manitoba, 1986.

Argues that American biologists, in rejecting both idealism and mechanism, turned towards organicist-materialism. Sees this third approach as allowing for a progressive evolution and the belief in free will and individualism. Believes that by the turn of the twentieth century, American biologists had synthesized physico-chemical approaches into their worldview. Relies heavily upon published articles and monographs.

397. Provine, William R. *Sewall Wright and Evolutionary Biology.*
    Chicago: University of Chicago Press, 1986. xvi+545
    pp. Appendix, Index.

    Provides a highly technical intellectual biography.
    Emphasizes Wright's contributions to evolutionary biology.
    Depends heavily upon interviews. Addresses the scientist as well
    as the historian.

CHAPTER VI: THE SOCIAL SCIENCES

GENERAL STUDIES

398. Bellomy, Donald C. "'Social Darwinism' Revisited." *Perspectives in American History,* n.s. 1 (1984): 1-129.

    Reviews the literature. Characterizes "Social Darwinism" as a derogatory phrase created on the Continent at the turn of the twentieth century which was appropriated by American sociologists. Argues that the confusion over the meaning of the term results from its negative connotation: the term was used to label ideas with which the user disagreed.

399. Bulmer, Martin. "Quantification and Chicago Social Science in the 1920s: A Neglected Tradition." *Journal of the History of the Behavioral Sciences,* 17 (1981): 312-331.

    Demonstrates that the significance of quantitative methods at Chicago has been neglected by historians. Blames this on the cross-disciplinary aspect of quantification. Focuses on the contributions to sociology and political science.

400. Cravens, Hamilton. "History of the Social Sciences." *Osiris*, 2nd ser., 1 (1985): 183-207.

    Argues that the social sciences are responses to questions of the relationships among human beings. Suggests that social sciences may be a misnomer; because of the emphasis on application, the fields could be called social technologies. Finds that the history of the social sciences is an underdeveloped and unbalanced field, with an extensive literature for some of the disciplines and little for others.

401. Cravens, Hamilton. "Recent Controversy in Human Develop-
      ment: A Historical View." *Human Development,* 30
      (1987): 325-335.

           Identifies three distinct ages in the history of the science
      of human development in the United States: child welfare (1870-
      1920), child development (1920-1950), and human development
      (since 1950). Argues that the prevalent worldview during the
      first ages was the inability of an individual to be thought of
      except in terms of group membership. Concludes that the
      increasing controversy in this science since 1950 is a result of the
      rejection of this deterministic view.

402. Fitzpatrick, Ellen Frances. "Academics and Activists: Women
      Social Scientists and the Impulse for Reform, 1892-1920."
      Ph.D. dissertation, Brandeis University, 1981.

           Examines a group of women--Edith Abbott, Sophonisba
      Breckinridge, Katherine B. David, and Frances Kellor--who
      studied the social sciences at the University of Chicago. Argues
      that these women successfully pursued their professors' views
      that social scientists should attempt both scientific understanding
      and social reform.

           Published as *Endless Crusade: Women Social Scientists
      and Progressive Reform.* New York: Oxford University Press,
      1990. xv+271 pp. Index.

403. Gillispie, Richard P. "Manufacturing Knowledge: A History of
      the   Hawthorne   Experiments."   Ph.D.   dissertation,
      University of Pennsylvania, 1985.

           Depicts the series of experiments on labor productivity,
      worker motivation, and the relations between managers and
      workers, at the Hawthorne Works of the Western Electric
      Corporation between 1924 and 1933. Characterizes the experi-
      ments, which led to a new approach to industrial management,
      as the principal application of social science research to this area.
      Sees the experiments as resulting from the need for a new
      ideology of management, the professionalization of management
      in America (the experiments were conducted in part by the
      faculty of the Harvard Business School), and the increasing
      interest of social scientists in developing a science of behavior
      and human control.

Published as *Manufacturing Knowledge: A History of the Hawthorne Experiments.* Cambridge: Cambridge University Press, 1991. x+282 pp. Bibliography, Index.

404. Leach, Eugene E. "Mastering the Crowd: Collective Behavior and Mass Society in American Social Thought, 1917-1939." *American Studies,* 27, No. 1 (Spring 1986): 99-113.

Credits American sociologists and psychologists with distinguishing between the mass, which is passive, and the crowd, which is active. Argues that Americans valued the mass because it could be easily directed by progressive elites.

405. Mabee, Carlton. "Margaret Mead and Behavioral Scientists in World War II: Problems in Responsibility, Truth, and Effectiveness." *Journal of the History of the Behavioral Sciences,* 23 (1987): 3-13.

Asks what are the ethical responsibilities of scientists. Argues that Mead overestimated the effectiveness of the social scientists in World War II.

406. Morawski, J. G. "Organizing Knowledge and Behavior at Yale's Institute of Human Relations." *Isis,* 77 (1986): 219-242.

Argues that the Institute of Human Relations failed to meet its objective of an integrated synthetic human science. Ascribes that failure to its inability to solve the problem of how imperfect humans study humans.

ANTHROPOLOGY AND ARCHAEOLOGY

407. Bieder, Robert E. "Anthropology and History of the American Indian." *American Quarterly,* 33 (1981): 309-326.

Contends that anthropologists, not historians, have shaped the historiography of the Native Americans because historians viewed Native Americans as obstacles to the progress of the United States and uninteresting in their own right. Detects a change in this attitude starting in the 1960s. Reviews the American anthropological theories of Native Americans.

408. Bieder, Robert E. *Science Encounters the Indian, 1820-1880: The Early Years of American Ethnology.* Norman: University of Oklahoma Press, 1986. xiii+290 pp. Bibliography, Index.

Explores the ideas of five ethnologists: Albert Gallatin, Samuel G. Morton, Ephraim George Squier, Henry Rowe Schoolcraft, and Lewis Henry Morgan. Finds that theories of social evolution change in response to new evidence and changing social views. Revises the material presented in Item I:633.

409. Cole, Douglas. *Captured Heritage: The Scramble for Northwest Coast Artifacts.* Seattle: University of Washington Press, 1985. xvi+373 pp. Index.

Describes the efforts by museums to collect Native American artifacts. Looks at the Smithsonian Institution under S. F. Baird, the work by Franz Boas for the American Museum of Natural History, and the work of Charles F. Newcombe for the Field Museum. Credits the declining interest in collecting at the turn of the twentieth century to a declining interest in artifacts in general and a surfeit of specimens in American museums. Sees the period of the effort to collect Northwest artifacts as coinciding with that of the great growth of American museums.

410. Darnell, Regna. "Personality and Culture: The Fate of the Sapirian Alternative." *Malinowski, Rivers, Benedict and Others: Essays on Culture and Personality* (item 439), pp. 156-183.

Argues that Sapir influenced others through personal interaction. Sees him as presenting an alternative to the dominant view that culture had formative power on the individual.

411. DeSimone, Alfred August, Jr. "Ancestors or Aberrants: Studies in the History of American Paleoanthropology, 1915-1940." Ph.D. dissertation, University of Massachusetts, 1986.

Studies the thought of five major figures in hominid evolution. Argues that this was a difficult period for American paleoanthropology. Sees a reluctance to acknowledge the known Pleistocene fossils as human ancestors, resulting in scenarios where there were no known ancestors of modern humans.

412. Dunnell, Robert C. "Five Decades of American Archaeology." *American Archaeology Past and Future: A Celebration of the Society for American Archaeology, 1935-1985* (item 434), pp. 23-49.

Identifies three major trends: a lessening of importance of amateurs, increasing employment in non-academic environments, and less basic research in the field.

413. Ebihara, May. "American Ethnology in the 1930s: Contexts and Currents." *Social Context of American Ethnology, 1840-1984* (item 421), pp. 101-121.

Identifies this as the formative period. Utilizes interviews with twenty-two anthropologists. Observes common features at all the major centers of graduate training. Lists faculty and doctorates at six major universities. Argues that the depression resulted in limited academic job opportunities and a general political shift to the left by the new Ph.D.s.

414. Fagette, Paul H., Jr. "Digging for Dollars: The Impact of the New Deal on the Professionalization of American Archaeology." Ph.D. dissertation, University of California, Riverside, 1985.

Argues that the increased opportunities for research provided by New Deal projects accelerated the development and professionalization of American archaeology. Concludes that the government support enabled archaeology to breakout from its earlier regionalization.

415. Fogelson, Raymond D. "Interpretations of the American Indian Psyche: Some Historical Notes." *Social Context of American Ethnology, 1840-1984* (item 421), pp. 4-27.

Reviews Euroamerican studies and impressions of Native American psychology. Sees a transformation of what had been originally viewed as positive traits into indicators of moral weakness or negative stereotypes. Concludes that antebellum American ethnology was little more than a summary of earlier popular imagery.

416. Fowler, Don D. "Conserving American Archaeological Resources." *American Archaeology Past and Future: A Celebration of the Society for American Archaeology, 1935-1985* (item 434), pp. 135-162.

Argues that despite a general interest in the results of archaeological research, Americans are indifferent to the need to preserve and protect archaeological sites, because Euroamericans do not see Native American artifacts as part of the American heritage. Contrasts this attitude with the conservation ethic of Mexico.

417. Frantz, Charles. "Relevance: American Ethnology and the Wider Society, 1900-1940." *Social Context of American Ethnology, 1840-1984* (item 421), pp. 83-100.

Argues that many American ethnologists active during this period thought that their discipline was relevant to the wider society and that the discipline was increasing its relevancy during the period. Finds two shifts in the focus of research during the period: from Native Americans to other peoples, and from historical and evolutionary to contemporary frameworks.

418. Geschwind, Norman. "The History of Anthropology and Education in the United States--From the Doctrine of Recapitulation to the Culture Concept: The Impact of a Paradigm Shift." Ed. D. dissertation, University of Hawaii, 1980.

Applies Thomas Kuhn's theory of paradigm shift to explain the history of the field of "anthropology and education" from the mid-nineteenth century to 1954. Focuses on G. Stanley Hall, Franz Boas, and Margaret Mead.

419. Gifford, John A., and George Rapp, Jr. "The Early Development of Archaeological Geology in North America." *Geologists and Ideas: A History of North American Geology* (item 282), pp. 409-421.

Argues that despite a lack of recognition of a formal subdiscipline of archaeological geology until 1977, there had been a tradition since about 1870 of geologists including archaeological field studies as part of their fieldwork. Identifies two major themes in the early work: the "early man" contro- versy, and the reconstruction of the physical environments of prehistoric humans.

420. Gruber, Jacob W. "Archaeology, History, and Culture." *American Archaeology Past and Future: A Celebration of the Society for American Archaeology, 1935-1985* (item 434), pp. 163-186.

Argues that until culture was envisioned as a complete and consistent system of behavior, archaeology could be little more than the collecting of artifacts.

421. Helm, June, editor. *Social Context of American Ethnology, 1840-1984*. Washington, D.C.: American Ethnological Society, 1985. vii + 184 pp.

Contains the proceedings of a meeting of the American Ethnological Society in 1984. Stresses the ideologies and issues that shaped American ethnology and ethnologists and the manner in which ethnology has been conducted. Includes items 413, 415, 417, 424, 429, 430, 444.

422. Hinsley, Curtis M., Jr. "Edgar Lee Hewett and the School of American Research in Santa Fe, 1906-1912." *American Archaeology Past and Future: A Celebration of the Society for American Archaeology, 1935-1985* (item 434), pp. 217-233.

Argues that American archaeology was more decentralized than other American sciences. Presents the founding of the School of American Research as a rebellion against the domination of the field by professionals in the Northeast. Raises a number of general and specific issues concerning the investigation of history by the professional historian as contrasted by the scientist acting as a historian.

423. Hinsley, Curtis M., Jr. "From Shell-heaps to Stelae: Early Anthropology at the Peabody Museum." *Objects and Others: Essays on Museums and Material Culture* (item 440), pp. 49-74.

Places the Peabody in the context of the transition of the study of humans from a humanistic field to a science and the need to reconcile archaeology as a part of human culture with evolutionary anthropology. Argues that evolution made museums give attention to the mundane. Concludes at the end of the nineteenth century with the shift of the Peabody's research focus to ancient Mayan culture.

424. Hinsley, Curtis M., Jr. "Hemispheric Hegemony in Early American Anthropology, 1841-1851: Reflections on John Lloyd Stephens and Lewis Henry Morgan." *Social Context of American Ethnology, 1840-1984* (item 421), pp. 28-40.

Sees this decade as a period both of great expansion and of societal splits along lines of wealth, gender, and race. Characterizes Stephen as rejecting contemporary Central American society in favor of the past ruins. Sees Morgan as an appropriator of Iroquois history for contemporary national purposes. Describes both men as voices of the emergent American ideology of inevitable economic and political hegemony over the hemisphere.

425. Hinsley, Curtis M., Jr. *Savages and Scientists: The Smithsonian Institution and the Development of American Anthropology, 1846-1910.* Washington, D.C.: Smithsonian Institution Press, 1981. 319 pp. Bibliography, Index.

Perceives the Smithsonian as dominating American anthropology--with a focus on Native Americans--until university departments appear at the turn of the century. Combines institutional, biographical, and intellectual history. Provides a detailed discussion of the thought of John Wesley Powell and a history of the Bureau of American Ethnology. This is a revision of Item I:638.

426. Horsman, Reginald. *Josiah Nott of Mobile: Southerner, Physician, and Racist Theorist.* Baton Rouge: Louisiana State University Press, 1987. xiii+347 pp. Index.

Explores the life and thought of an ethnologist who defended white supremacy on scientific grounds and the freedom of science from religious restriction. Finds that Nott's anthropology did not depend on research but on his prejudicial reading of the work of others and history.

427. Jacknis, Ira. "Franz Boas and Exhibits: On the Limitations of the Museum Method of Anthropology." *Objects and Others: Essays on Museums and Material Culture* (item 440), pp. 75-111.

Discusses Boas's activities at the American Museum of Natural History, 1895-1905. Argues that Boas saw museums as serving three distinct purposes: entertainment, instruction, and research. Uses the Hall of Northwest Coast Indians as a case study. Analyzes the exhibition in detail, including hall arrangement, installation, labels and text. Concludes that Boas left the Museum because he came to recognize that exhibitions had limitations as communicators and that his interests had shifted from artifact-based research to more observational and behavioral anthropology.

428. Jackson, Walter. "Melville Herskovits and the Search for Afro-American Culture." *Malinowski, Rivers, Benedict and Others: Essays on Culture and Personality* (item 439), pp. 95-126.

    Credits Herskovits's changing views in the late 1920s and 1930s concerning the retention of African elements in African-American culture to his fieldwork in Surinam and Dahomey, which resulted in evidence of African influence on African-American culture.

429. Kehoe, Alice B. "The Ideological Paradigm in Traditional American Ethnology." *Social Context of American Ethnology, 1840-1984* (item 421), pp. 41-49.

    Observes that American ethnology was shaped by larger ideological metaphors reflecting and supporting the economic and political structure. Sees this very clearly in the writings of John Wesley Powell and the staff of the Bureau of American Ethnology. Argues that Powell's internalized factors important to the capitalist enterprise into his scientific practice.

430. Kelly, Lawrence C. "Why Applied Anthropology Developed When It Did: A Commentary on People, Money, and Changing Times, 1930-1945." *Social Context of American Ethnology, 1840-1984* (item 421), pp. 122-138.

    Addresses issues usually addressed by practitioners' history. Agrees with the main conclusions of the practitioners. Credits the impetus for applied anthropology during the New Deal to government administrators, not academics. Provides an intellectual genealogy of the early practitioners of applied anthropology in the United States.

431. McMillan, Robert Lee. "The Study of Anthropology, 1931 to 1937, at Columbia University and the University of Chicago." Ph.D. dissertation, York University, 1986.

    Tests and rejects Leslie White's model of social organization as an explanation of the programs of these two institutions. Concludes that there no specific schools of thought at these schools.

432. Mark, Joan. "Francis La Flesche: The American Indian as Anthropologist." *Isis*, 73 (1982): 497-510.

Finds that it took thirty years for La Flesche to gain recognition as an anthropologist rather than just an informant. Concludes that Alice Fletcher used La Flesche to promote herself rather than assist his career.

433. Mark, Joan. *Four Anthropologists: An American Science in Its Early Years.* New York: Science History Publications, 1980. viii+209 pp. Bibliography, Index.

Examines the careers and work of Frederic Ward Putnam, Alice Cunningham Fletcher, Frank Hamilton Cushing, and William Henry Holmes to demonstrate that the United States had a strong presence in anthropology in 1900. Rejects the idea that Franz Boas was responsible for the creation of American anthropology. Argues that he joined an already flourishing discipline.

434. Meltzer, David J., Don D. Fowler, and Jeremy A. Sabloff, editors. *American Archaeology Past and Future: A Celebration of the Society for American Archaeology, 1935-1985.* Washington, D.C.: Smithsonian Institution Press, 1986, 479 pp. Bibliography.

Contains both historical articles and reviews of the state of the discipline. Includes items 412, 416, 420, 422, 443.

435. Meltzer, David J. "The Antiquity of Man and the Development of American Archaeology." *Advances in Archaeological Method and Theory,* 6 (1983): 1-51.

Reviews the history of the early man controversy from 1860 to 1920, concentrating on the period 1890-1927. Concludes that the archaeology practiced by the Bureau of American Ethnology was rooted in the assumption that the historic Native Americans and the prehistoric inhabitants of North America were one and the same.

436. Modell, Judith Schachter. *Ruth Benedict: Patterns of a Life.* Philadelphia: University of Pennsylvania Press, 1983. x+355 pp. Bibliography, Index.

Offers a thematic rather than strictly chronological biography and analysis of the development of Benedict's anthropology. Affirms that patterns are the key to understanding Benedict's life. Avoids value judgments on controversial aspects of Benedict's life.

437. Porter, Joseph C. *Paper Medicine Man: John Gregory Bourke and His American West.* Norman: University of Oklahoma Press, 1986. xix+362 pp. Bibliography, Index.

Examines the life and ideas of an army officer who became a chronicler of the Apache and an advocate of Indian rights. Describes Bourke's perceived mission as integrating Indians into White culture. Characterizes Bourke as an advocate of Indian economic self-sufficiency.

438. Reed, James Steven. "Clark Wissler: A Forgotten Influence in American Anthropology." Ph.D. dissertation, Ball State University, 1980.

Attempts to rescue Wissler, who was a curator at the American Museum of Natural History, from obscurity. Argues that the impact of a given individual on a discipline, both directly and indirectly, is governed by three factors: patronage, both received and given; the leaders, both intellectually, and institutionally, that one associates with; and the number and nature of one's disciples. Shows that Wissler, who had an extensive institutional network during his lifetime, but few students, was influential during his career, but was quickly forgotten.

439. Stocking, George W., Jr. editor. *Malinowski, Rivers, Benedict and Others: Essays on Culture and Personality.* Madison: University of Wisconsin Press, 1986, viii+257 pp. Index.

Focuses on the influence of psychological theory on anthropologists between the World Wars. Argues that the study of the relationship between personality and culture was a major force in American anthropology after World War I, but was non-existent by the 1960s. Includes items 90, 410, 428, 445.

440. Stocking, George W., Jr., editor. *Objects and Others: Essays on Museums and Material Culture.* Madison: University of Wisconsin Press, 1985. ix+285 pp. Index.

Focuses on the Museum Period in the history of anthropology, roughly 1860-1930. Includes items 105, 134, 423, 427.

441. Stocking, George W., Jr. "The Santa Fe Style in American Anthropology: Regional Interest, Academic Initiative, and Philanthropic Policy in the First Two Decades of the Laboratory of Anthropology, Inc." *Journal of the History of the Behavioral Sciences,* 18 (1982): 3-19.

Describes the Laboratory of Anthropology as modelled on the Marine Biological Laboratory in Woods Hole; financial support came from John D. Rockefeller, Jr.

442. Trautmann, Thomas R. *Lewis Henry Morgan and the Invention of Kinship.* Berkeley: University of California Press, 1987. xv + 290 pp. Bibliography, Index.

Studies the origin and evolution of Morgan's *Systems of Consanguinity and Affinity in the Human Family* (1871). Places the writing of the book in the context of the ethnological time revolution resulting from the discovery of fossil man. Demonstrates the role of Joseph Henry in the construction of the final form of the book.

443. Trigger, Bruce G. "Prehistoric Archaeology and American Society." *American Archaeology Past and Future: A Celebration of the Society for American Archaeology, 1935-1985* (item 434) pp. 187-215.

Addresses the role of the influence of the values of the American middle class on American archaeology. Argues that archaeological facts can force the adjustment of general beliefs if the intellectual climate is favorable.

444. Washburn, Wilcomb E. "Ethical Perspectives in North American Ethnology." *Social Context of American Ethnology, 1840-1984* (item 421), pp. 50-64.

Addresses two different forms of ethical behavior: responsibility to the advancement of knowledge and respect of the integrity of the people being studied. Contrasts Frank Hamilton Cushing's respect for the native society, but theft of artifacts, with Sol Tax's action anthropology, which respected the possessions of the people, but sought to reshape the structure of the subject's society.

445. Yans-McLaughlin, Virginia. "Science, Democracy, and Ethics: Moralizing Culture and Personality for World War II." *Malinowski, Rivers, Benedict and Others: Essays on Culture and Personality* (item 439), pp. 184-217.

Focuses on the attempts by Margaret Mead and Gregory Bateson to utilize anthropology for the Allied effort during World War II. Describes her attempts to help morale on the home front and his propagandizing of the enemy. Concludes that they had made the pragmatic decision that democracy was a prior and necessary condition for scientific detachment.

GEOGRAPHY

446. Blouet, Brian W., editor. *The Origins of Academic Geography in the United States.* Hamden, Conn.: Archon Books, 1981. xii+342 pp. Appendix.

    Contains twenty papers addressing the institutional and intellectual history of geography. Includes item 447.

447. Mikesell, Marvin W. "Continuity and Change." *The Origins of Academic Geography in the United States* (item 446), pp. 1-15.

    Argues that continuity has been the dominant theme of American geography. Identifies the most important change to have been the migration of American academic geography from the physical sciences, especially geology, to the social sciences.

PSYCHOLOGY

448. Bjork, Daniel W. *The Compromised Scientist: William James in the Development of American Psychology.* New York: Columbia University Press, 1983. xiv+221 pp. Bibliography, Index.

    Explains James's emergence as an experimental psychologist in terms of compromises between extremes. Views James as part of a transitional generation of intellectuals responding to a period of crisis and innovation.

449. Blight, James G. "Jonathan Edward's Theory of the Mind: Its Applications and Implications." *Explorations in the History of Psychology in the United States* (item 451), pp. 61-120.

    Assesses Edward's theory that we cannot know with certainty; human reason is fallible. Suggests that the differences between Edward's physiological theory and that of William James are not that great. Includes a historiographical review of Edward's psychological thought. Argues that there is a bias in the literature against pre-experimental psychology.

450. Brožek, Josef. "David Jayne Hill: Between the Old and New Psychology." *Explorations in the History of Psychology in the United States* (item 451), pp. 121-147.

     Surveys the life of the first professor of psychology in the United States. Sees Hill, who taught at the University at Lewisburg (now Bucknell University), as a transition figure.

451. Brožek, Josef, editor. *Explorations in the History of Psychology in the United States.* Lewisburg: Bucknell University Press, 1984. 333 pp. Index.

     Provides a forum for monographic studies which fall between journal articles and books. Includes items 449, 450, 456, 461, 467, 473.

452. Buckley, Kerry W. "Behaviorism and the Professionalization of American Psychology: A Study of John Broadus Watson, 1878-1958." Ph.D. dissertation, University of Massachusetts, 1982.

     Sees Watson as part of the generation of professionals who found opportunities in solving the problems created by industrialization. Describes Watson as the first "pop" psychologist, appealing to the expanding middle class. Describes Watson's vision of behaviorism as a way of directing human activity into pre-determined courses.

     Published as *Mechanical Man: John Broadus Watson and the Beginnings of Behaviorism.* xv + 233 pp. Index. New York: Guilford Press, 1989.

453. Buckley, Kerry W. "The Selling of a Psychologist: John Broadus Watson and the Application of Behavioral Techniques to Advertising." *Journal of the History of the Behavioral Sciences,* 18 (1982): 207-221.

     Looks at the contributions of Watson to the development of market research techniques while he was at the J. Walter Thompson Advertising Company.

454. Capshew, James Herbert. "Psychology on the March: American Psychologists and World War II." Ph.D. dissertation, University of Pennsylvania, 1986.

     Argues that World War II served as a catalyst for the professionalization of applied psychology. Describes a wide range of uses of psychologists in the war effort. Concludes that psychology was transformed from a largely academic discipline to a consulting profession, offering practical information.

455. Chapman, Paul Davis. "Schools as Sorters: Lewis M. Terman and the Intelligence Testing Movement, 1890-1930." Ph.D. dissertation, Stanford University, 1980.

Identifies three reasons why intelligence testing was adopted rapidly: support by psychologists and public school administrators; the need to efficiently and effectively classify students as a response to problems facing public schools around 1900; and tests fit into the value system of the Progressive Era. Provides three case studies of California school systems. Selects Terman as a focus because he was a leader in the movement, especially in the 1920s.

Published as *Schools as Sorters: Lewis M. Terman, Applied Psychology, and the Intelligence Testing Movement, 1890-1930.* New York: New York University Press, 1988. xv+228 pp. Bibliography, Index.

456. Evans, Rand B. "The Origins of American Academic Psychology." *Explorations in the History of Psychology in the United States* (item 451), pp. 17-60.

Rejects the theory that psychology was created in the 1880s. Advocates the interpretation that this period marked its institutionalization as an experimental discipline. Defines psychology as the study of how one knows, feels, or does. Traces the field to Harvard College in the late seventeenth century.

457. Henle, Mary. "Robert M. Ogden and Gestalt Psychology in America." *Journal of the History of the Behavioral Sciences,* 20 (1984): 9-19.

Details Ogden's contributions in introducing Gestalt psychology to the United States.

458. Hilgard, Ernest R. *Psychology in America: A Historical Survey.* San Diego: Harcourt Brace Jovanovich, 1987. xxix+1009 pp. Bibliography, Indices.

Organized around basic and applied research areas. Stresses theories and experiments, but also considers personalities, institutional settings, and social factors.

459. Minton, Henry L. "Lewis M. Terman and Mental Testing: 'In Search of the Democratic Ideal'." *Psychological Testing and American Society, 1890-1930* (item 470), pp. 95-112.

   Describes Terman's efforts to reconcile the American democratic ideal that there were no inborn differences in mental functions among citizens with the scientific data which contradicted the ideal.

460. O'Donnell, John M. *The Origins of Behaviorism: American Psychology, 1870-1920.* New York: New York University Press, 1985. xii+299pp. Index.

   Highlights the connections between the new psychology--which he sees as evolving from a mixture of philosophy, physiology, and social science--and the old moral philosophy. Stresses that psychology was viewed as an empirical science which would serve to reinforce traditional social and philosophical beliefs and values. Presents behaviorism as a form of experimental psychology which enabled its practitioners to separate and distinguish psychology from philosophy and physiology, while still promising practical contributions to society.

461. Popplestone, John A., and Marion White McPherson. "Pioneer Psychology Laboratories in Clinical Settings." *Explorations in the History of Psychology in the United States* (item 451), pp. 196-272.

   Uses published research reports during the period 1890-1910. Organizes the material by individual researchers. Describes the experimental procedures as non-uniform and frequently sloppy. Assesses these laboratories as a response to the self-doubts of psychology about its scientific credentials.

462. Reed, James. "Robert M. Yerkes and the Mental Testing Movement." *Psychological Testing and American Society, 1890-1930* (item 470), pp. 75-94.

   Argues that Yerkes's work in Army testing contradicted his earlier experimentation. Describes him as a biologically oriented psychologist interested in charting the region between mind and body.

463. Samelson, Franz. "Struggle for Scientific Authority: The Reception of Watson's Behaviorism, 1913-20." *Journal of the History of the Behavioral Sciences*, 17 (1981): 399-425.

Argues that there was considerable resistance to behaviorism because acceptance required that psychologists change their perception of reality.

464. Samelson, Franz. "Was Early Mental Testing (a) Racist Inspired, (b) Objective Science, (c) A Technology for Democracy, (d) The Origin of Multiple-Choice Exams, (e) None of the Above? (Mark the RIGHT Answer)." *Psychological Testing and American Society, 1890-1930* (item 470), pp. 113-127.

Credits the invention of the multiple-choice question to Frederick J. Kelly, who developed it in the search for a more efficient and effective reading test.

465. Scarborough, Elizabeth, and Laurel Furumoto. *Untold Lives: The First Generation of American Women Psychologists*. New York: Columbia University Press, 1987. xiv+236 pp. Appendices, Bibliography, Index.

Believes women have been ignored in historical accounts of psychology. Illustrates the themes of the denial of opportunity and the conflict between family claims and career through the lives of five individuals. Provides a collective portrayal of twenty-five additional female psychologists active by 1906.

466. Sokal, Michael M. "The Gestalt Psychologists in Behaviorist America." *American Historical Review*, 89 (1984): 1240-1263.

Rejects the view that the diffusion of Gestalt psychology in the United States was a result of the migration of German psychologists fleeing Hitler. Concludes that Gestalt psychology was firmly established in the United States by 1930 and that the German emigrants were well received.

467. Sokal, Michael M. "James McKeen Cattell and American Psychology in the 1920s." *Explorations in the History of Psychology in the United States* (item 451), pp. 273-323.

Reminds us that most psychologists came from the native-born, white, Protestant middle class and shared many of the values and biases of that group. Assesses the psychological community during the period 1918-1929. Describes the period as one of great confidence in the ability of psychologists to solve problems. Believes that this confidence led psychologists to overstate their claims and in turn led to exaggerated expectations on the part of the public. Calls for additional work in the social history of psychology.

468. Sokal, Michael M. "James McKeen Cattell and the Failure of Anthropometric Mental Testing, 1890-1901." *The Problematic Science: Psychology in Nineteenth-Century Thought.* Edited by William R. Woodward and Mitchell G. Ash. New York: Praeger, 1982, pp. 322-345.

Views Cattell as very influential on American culture, although not a great scientist. Argues that Cattell's scientific research was based on out-of-date methods. Characterizes it as a combination of Baconian empirical fact gathering with Comtean appreciation for mathematics. Stresses that Cattell assumed that quantifiable data was the only type worthy of scientific attention.

469. Sokal, Michael M. "The Origins of the Psychological Corporation." *Journal of the History of the Behavioral Sciences,* 17 (1981): 54-67.

Describes the early history of a venture in applied psychology founded by James McKeen Cattell. Concludes that the Corporation was an example of Cattell's effort to practice nineteenth-century science in the twentieth century. Sees it as part of the general movement by science in the 1920s to shape society by applying scientific knowledge.

470. Sokal, Michael M., editor. *Psychological Testing and American Society, 1890-1930.* New Brunswick: Rutgers University Press, 1987. xiii+205 pp. Index.

Contains seven papers, most taking a biographical approach. Includes items 200, 459, 462, 464, 471.

471. Von Mayrhauser, Richard T. "The Manager, the Medic, and the Mediator: The Clash of Professional Psychological Styles and the Wartime Origins of Group Mental Testing." *Psychological Testing and American Society, 1890-1930* (item 470), pp. 128-157.

Describes the split between Robert M. Yerkes, president of the American Psychological Association, and Walter D. Scott,

America's leading applied psychologist. Compares Yerkes's desire to do medical style of testing to diagnose mental incompetence with Scott's experience using mass testing to identify specific human qualities leading to accomplishment in specific areas.

472. Windholz, George. "Pavlov's Position toward American Behaviorism." *Journal of the History of the Behavioral Sciences,* 19 (1983): 394-407.

Argues that I. P. Pavlov saw American behaviorism as a confirmation of his theory and method.

473. Woodward, William R. "William James's Psychology of Will: Its Revolutionary Impact on American Psychology." *Explorations in the History of Psychology in the United States* (item 451), pp. 148-195.

Argues that James's reworking of European models provided the basis for the development of the formulations which characterize American psychology. Sees the psychological explanation of freedom as still the fundamental issue, despite changes in terminology over the years.

SOCIOLOGY

474. Bannister, Robert C. *Sociology and Scientism: The American Quest for Objectivity, 1880-1940.* Chapel Hill: University of North Carolina Press, 1987. viii+301 pp. Index.

Views the quest by sociologists for objectivity as an effort to provide compelling arguments or justifications for prescriptive judgments and practical reforms. Sees the sociologists as embodying a secularized Protestant spirit.

475. Barber, Barnard. "Theory and Fact in the Work of Talcott Parsons." *The Nationalization of the Social Sciences* (item 217), pp. 123-139.

Ascribes Parsons's tendency to allow theoretical inspiration to exceed its empirical base to his strength as a theoretician, his lack of awareness of the measurement problem, and his lack of involvement in empirical research on a sustained basis.

476. Bierstedt, Robert. *American Sociological Theory: A Critical History.* New York: Academic Press, 1981. xiv+525 pp. Appendices, Index.

Examines the lives and theories of ten major sociologists from William Graham Sumner through Robert Merton. Views sociological theory as constructed within broader philosophical issues. Provides a very negative analysis for those sociologists he disagrees with.

477. Bulmer, Martin. *The Chicago School of Sociology: Institutionalization, Diversity, and the Rise of Sociological Research.* Chicago: University of Chicago Press, 1984. xix + 285 pp. Index.

Argues that diversity and an interdisciplinary orientation were the chief characteristics of the Chicago School. Claims that other historians have exaggerated the influence of George Herbert Mead.

478. Howard, Ronald L. *A Social History of American Family Sociology, 1865-1940.* Edited by John Mogey. Westport: Greenwood, 1981. xv + 150 pp. Index.

Studies the second largest specialty within sociology, focusing on textbooks (pre-1920) or journals (post-1920) as indicators of the content of the field. Depends upon Stephen Toulmin's work for theoretical underpinning. Traces the evolution of the field from a forum for moral debate to an empirical science.

479. Johnston, Barry V. "Pitrim Sorokin and the American Sociological Association: The Politics of a Professional Society." *Journal of the History of the Behavioral Sciences,* 23 (1987): 103-122.

Discusses the 1963 campaign electing Sorokin president of the American Sociological Association by write-in. Places the election in the context of the debate over whether the presidency of professional societies is an honorific post, used to acknowledge past achievement, or an administrative one, reserved for those who can provide leadership.

480. Kuklick, Henrike. "Boundary Maintenance in American Sociology: Limitation to Academic 'Professionalization'." *Journal of the History of the Behavioral Sciences,* 16 (1980): 201-219.

Uses the history of sociology to test the professionalization model which sociologists developed. Finds that it does not fit.

Argues that institutionalization of a field does not necessarily lead to consensus on the content of that field.

481. Lewis, J. David, and Richard L. Smith. *American Sociology and Pragmatism: Mead, Chicago Sociology, and Symbolic Interaction.* Chicago: University of Chicago Press, 1980. xx + 356 pp. Appendices, Index.

Challenges the accepted historiography in both the history of American philosophy and the history of sociology. Argues that there were two forms of pragmatism--nominalism, represented by the thought of William James and John Dewey, and realism--represented by Charles Sanders Peirce and George Herbert Mead. Claims that Mead was peripheral to early University of Chicago sociology. The first half of the book is highly internal history of philosophy. The second half depends upon surveys of former students in sociology at Chicago.

482. Lowy, Richard Frank. "George Herbert Mead: Evolutionary Naturalism, the Act, Science, and Social Reform." Ph.D. dissertation, University of California, Riverside, 1984.

Argues that Mead developed his ideas within a naturalistic framework, emphasizing the findings of evolutionary biology and relativity physics. Contains an extensive secondary bibliography on Mead.

CHAPTER VII: AGRICULTURE AND TECHNOLOGY

AERONAUTICS AND ASTRONAUTICS

483. Bilstein, Roger E. *Flight in America, 1900-1983: From the Wrights to the Astronauts.* Baltimore: Johns Hopkins University Press, 1984. xx + 356 pp. Bibliography, Index.

   Surveys the technical, social, economic, and political aspects of aerospace history. Stresses general aviation more than most general histories of aviation. Treats the interaction of flight and culture very superficially.

484. Bilstein, Roger E. *Flight Patterns: Trends of Aeronautical Development in the United States, 1918-1929.* Athens: University of Georgia Press, 1983. xi + 236 pp. Bibliography, Index.

   Views this period as the end of the pioneering phase in aviation technology. Provides new information garnered from extensive research in manuscript and printed sources.

485. Connaughton, Michael E. "'Ballomania', the American Philosophical Society and Eighteenth-Century Science." *Journal of American Culture,* 7, No. 1 and 2 (Spring/Summer 1984): 71-74.

   Argues that the American Philosophical Society refused to endorse manned balloon flight in the 1780s because of its fear of being associated with a controversial activity.

486. Corn, Joseph J. *The Winged Gospel: America's Romance with Aviation, 1900-1950.* New York: Oxford University Press, 1983. xii + 177 pp. Index.

Argues that the airplane became a symbol of the promise of the future--peace and prosperity through technology. Aviation had been raised to the status of a technological religion, having assimilated the metaphors and symbols of the spiritual life. Discusses the role of women pilots and the education of the youth through model airplanes and flying clubs in the spread of the gospel.

487. Crouch, Tom D. *A Dream of Wings: Americans and the Airplane, 1875-1905.* New York: W. W. Norton & Company, 1981. 349 pp. Bibliography, Index.

Focuses on the activities of Octave Chanute, Samuel P. Langley, and the Wrights. Views the Wright brothers as the culmination of a generation of research by a community of scientists and engineers. Credits the Wrights' success to their recognition that an aircraft was a complex technical system.

488. Crouch, Tom D. *The Eagle Aloft: Two Centuries of the Balloon in America.* Washington, D.C.: Smithsonian Institution Press, 1983. 770 pp. Bibliography, Index.

Provides a definitive account of ballooning ascents in the United States. Contends that the technology of gas balloons became stable very quickly; for most of the two centuries technological changes have been in the form of increasing scale. Places the technology within social and biographical contexts. Provides only a very limited discussion of balloons during World War I.

489. Hacker, Barton C. "Robert H. Goddard and the Origins of Space Flight." *Technology in America: A History of Individuals and Ideas* (item 073), pp. 228-240.

Views space flight as more of an engineering problem, ultimately dependent upon a large-scale organization, than a technological one. Sees Goddard as contributing to the theory of space flight, but having little direct role in its achievement.

490. Hallion, Richard P. "A Source Guide to the History of Aeronautics and Astronautics." *American Studies International,* 20 (1982): 3-50.

Includes a critical essay covering the history of these fields by historical periods and a bibliography. Emphasizes

American contributions and events, although it includes non-American subjects as well.

491. Hanle, Paul A. *Bringing Aerodynamics to America.* Cambridge: The M.I.T. Press, 1982. xiv+184 pp. Index.

Explores the role of Theodore von Kármán in the rise of American expertise in theoretical aerodynamics during the 1920s and 1930s. Credits the Guggenheim Fund, the result of the aircraft industry's recognition of the scientific basis of the industry, with supplying the major financial wherewithal. Demonstrates the dependency of science upon the interplay between personalities, intellectual conditions, and institutional circumstances.

492. Horwitch, Mel. *Clipped Wings: The American SST Conflict.* Cambridge: MIT Press, 1982. x+473 pp. Index.

Approaches the SST conflict as a battle over how to develop new technology and the control over decisions. Sees an increasing societal involvement in technological issues. Concludes that the defenders of the SST lacked management skills, but they also lacked a significant justification for the new technology.

AGRICULTURE

493. Daniel, Pete. *Breaking the Land: The Transformation of Cotton, Tobacco, and Rice Cultures since 1880.* Urbana: University of Illinois Press, 1985. xvi+352 pp. Index.

Examines the cultures and work cycles centered upon these commodities. Finds that technological innovation and governmental programs destroyed the traditional cotton culture by the end of World War II. Observes that after 1941 both the tobacco and rice cultures were also transformed. Concludes that in all three cultures the trend has been to fewer producers and larger holdings.

494. Dreyer, Peter. *A Gardener Touched with Genius: The Life of Luther Burbank.* Berkeley: University of California Press, 1985. Revised ed. xiii+293 pp. Appendices, Index.

Corrects past errors and distortions in the historical literature. Sees Burbank as a follower of the neo-Lamarkian Darwin of *Variation.* Credits Burbank with arousing interest in horticulture. Argues that his approach to plants more closely resembled the work of an artist than that of a scientist.

495. Fitzgerald, Deborah K. "The Business of Breeding: Public and
      Private Development of Hybrid Corn in Illinois, 1890-
      1940." Ph.D. dissertation, University of Pennsylvania,
      1985.

      Provides a case study for two questions: what is the
      process by which pure science becomes applied science; and
      how does this process differ between land-grant institutions and
      private industry. Sees the Illinois Agricultural Experiment Station
      and seed companies following parallel paths, and even cooper-
      ating, as they attempted to improve field performance and yield
      through the application of scientific principles. Finds a diver-
      gence of interests, however, when hybrids ceased being
      experimental. Concludes that the aggressive marketing by the
      seed companies in the 1930s forced the Experiment Station out
      of hybrid seed development.

      Published as *The Business of Breeding: Hybrid Corn in
      Illinois, 1890-1940*. Ithaca: Cornell University Press, 1990.
      xiv + 247 pp. Bibliography, Index.

496. Heitmann, John A. *The Modernization of the Louisiana Sugar
      Industry, 1830-1910*. Baton Rouge: Louisiana State
      University Press, 1987. xii + 298 pp. Bibliography, Index.

      Utilizes the institution--trade organization, experiment
      station, and university--as the key unit in analysis. Argues that
      the institutions were created to provide stability and control in an
      industry threatened by foreign competition and problems of a
      new labor system. Sees science and technology as important
      elements in the response of the sugar planters to changing
      conditions. Assesses the mutually important relationship between
      the Louisiana Sugar Planters Association, which recognized both
      the need for science/technology and Louisiana's lack of chemists
      and engineers, and the United States Department of Agriculture,
      which wanted to make the nation self-sufficient in the production
      of sugar. Describes the research conducted at the Audubon Sugar
      School, which became a major center for carbohydrate
      chemistry.

497. Hurt, R. Douglas. *American Farm Tools: From Hand Power to
      Steam Power*. Manhattan, Kansas: Sunflower University
      Press, 1982. 121 pp. Appendix, Bibliography, Index.

      Focuses on the basic tools for the production of cereal
      grains, hay and fodder from the colonial period to World War
      I. Finds that technological advance in one stage of farming has

required change in another. Claims that by the mid-nineteenth century a technological balance had been achieved. Well-illustrated, but lacks footnotes.

498. Marcus, Alan I. *Agricultural Science and the Quest for Legitimacy: Farmers, Agricultural Colleges, and Experiment Stations, 1870-1890.* Ames: Iowa State University Press, 1985. x+269 pp. Bibliography, Index.

Studies the context and passage of the Hatch Act in the hope of providing more general insight into the procedure followed by groups and professions seeking legitimacy. Finds two conflicting views as to the solution to the problems facing American farmers during the last third of the nineteenth-century: either farmers had to approach agriculture as a business and develop the proper system, or as a profession based on scientific information. Views the Hatch Act as a victory for systematic farmers over scientific agriculturalists.

499. Pursell, Carroll W., Jr. "Cyrus Hall McCormick and the Mechanization of Agriculture." *Technology in America: A History of Individuals and Ideas* (item 073), pp. 71-79.

Observes that mechanization was important in the nineteenth century only for large farms and/or certain crops. States that the mass production of and improvements in hand tools were more significant for most farmers.

500. Rothstein, Morton. "Technological Change and American Farm Movements." *Technology, the Economy, and Society* (item 047), pp. 186-222.

Looks at two forms of farm movements: protest and self-help. Sees the former as sometimes targeting technology. Describes the latter's efforts to disseminate information.

501. Smith, David C. *The Maine Agricultural Experiment Station: A Bountiful Alliance of Science and Husbandry.* Orono: Maine Life Sciences and Agricultural Station, University of Maine at Orono, 1980. xiv+292 pp. Appendices, Index.

Provides an administrative history organized around directors. Demonstrates the tensions between basic research, applications, and administration. Links the decline of the station's scientific work to the decline of Maine's national importance in agriculture.

502. Williams, Robert C. *Fordson, Farmall, and Poppin' Johnny: A History of the Farm Tractor and Its Impact on America.* Urbana: University of Illinois Press, 1987. ix+232 pp. Bibliography, Index.

Provides an excellent technological history of the tractor and analysis of the tractor's impact on American farming and society. Sees the tractor as the major factor in overproduction by American farmers.

503. Wines, Richard A. *Fertilizer in America: From Waste Recycling to Resource Exploitation.* Philadelphia: Temple University Press, 1985. vii+247 pp. Appendix, Bibliography, Index.

Treats fertilizer as a technological system which evolved in an effort to maintain stability. Identifies the method of maintaining stability as small substitutions in the system, although these substitutions might ultimately lead to radical changes. Finds that fertilizer was part of an urban-rural recycling system during the first half of the nineteenth century. Argues that the introduction of guano was the first step in the transition to commercial fertilizers, resulting in a much more complex technological system.

504. Wines, Richard A. "The Nineteenth-Century Agricultural Transition in an Eastern Long Island Community." *Agricultural History,* 55 (1981): 50-63.

Finds little evidence of cultural resistance to technological change. Credits the rejection of inventions to the lack of suitability for local growing conditions.

505. Worster, Donald. "Hydraulic Society in California: An Ecological Interpretation." *Agricultural History,* 56 (1982): 503-515.

Maintains that the role of irrigation in California agriculture transformed it into a hydraulic society. Argues that the domination of one people by another is necessary in such societies. Contrasts California agricultural, where the farmers are dominated by impersonal capitalist, corporate pressures, to the ancient Nile River Valley society, where control was by an individual personal ruler. Warns of the danger that the environment might rebel against continued irrigation.

CHEMICAL TECHNOLOGY

506. Hochheiser, Sheldon. *Rohm and Haas: History of a Chemical Company.* Philadelphia: University of Pennsylvania Press, 1986. xiii+231 pp. Appendices, Index.

Summarizes the company strategy as staying at the frontier of science and technology, avoiding consumer goods, selling to industrial customers, and working with those customers to solve problems. Although this is a company-sponsored history to mark the company's seventy-fifth anniversary, it is well-documented and objective.

507. Meikle, Jeffrey L. "Materials and Metaphors: Plastics in American Culture." *New Perspectives on Technology and American Culture* (item 075), pp. 31-47.

Reviews the conceptions of plastic expressed by the industry and the applications of the material. Finds that in the early twentieth century plastic was a symbol of status and of the potential of American technology. After World War II, plastic became the symbol of change, transformation, and the desire to shape and control the environment.

508. Meikle, Jeffrey L. "Plastic, Material of a Thousand Uses." *Imagining Tomorrow: History, Technology, and the American Future* (item 049), pp. 77-96.

Finds that the negative effects of plastic were recognized only after some of the utopian dreams had actually become a reality.

509. Multhauf, Robert P. "Potash." *Material Culture of the Wooden Age* (item 056), pp. 227-240.

Links chemical ideas of the composition of potash to the practical techniques of production.

510. Reynolds, Terry S. "Defining Professional Boundaries: Chemical Engineering in the Early 20th Century." *Technology and Culture,* 27 (1986): 694-716.

Finds that establishing a boundary between chemical engineering and chemistry was difficult because chemists opposed a separate engineering discipline. Argues that the discipline eventually arose because the need of the American chemical industry for individuals with engineering skills, the concern over the role and status of chemists within industry, and the perceived threat from other engineering disciplines.

511. Reynolds, Terry S. *75 Years of Progress--a History of the American Institute of Chemical Engineers, 1908-1983.* New York: American Institute of Chemical Engineers, 1983. ii+ 200 pp. Appendix.

    Argues that the discipline of chemical engineering arose in the United States because of its uniquely high volume chemical industry and the resulting need to combine engineering and chemical skills. Sees the early years of the society as a struggle to establish the legitimacy of the discipline of chemical engineering. Defines legitimacy in terms of education: other engineering disciplines had an academic structure before the establishment of professional societies, but the chemical engineers reversed the process. This was the first engineering society to accredit curriculum. Credits the concept of "unit operations" with giving chemical engineering an identity. Includes anecdotal remarks by past presidents.

512. Sturchio, Jeffrey Louis. "Chemists and Industry in Modern America: Studies in the Historical Application of Science Indicators." Ph.D. dissertation, University of Pennsylvania, 1981.

    Employs the methods of science indicators to analyze major trends. Looks at the changing roles of chemists in industry since 1876, especially the shift from consulting to corporate chemist and the rise of industrial research laboratories. Provides two micro-analyses: the changing organization of research and development at the Du Pont Corporation, and the status of industrial chemists within the American Chemical Society.

513. Sturchio, Jeffrey L. editor. *Corporate History and the Chemical Industries: A Resource Guide.* Philadelphia: Center for History of Chemistry, 1985. 53 pp.

    Includes annotated bibliographies of over one hundred secondary sources, material on archives and records management, and discussions of oral history.

514. Taylor, Graham D. and Patricia E. Sudnik. *Du Pont and the International Chemical Industry.* Boston: Twayne Publishers, 1984. xxi+251 pp. Bibliography, Index.

Reviews the history of the Du Pont Corporation. Discusses the rise and fall of international cartels, consolidated, vertically integrated corporations, and foreign investment in the twentieth century. Focuses on the impact of government policies, including anti-trust suits, on both the company and the industry as a whole.

515. Trescott, Martha Moore. *The Rise of the American Electro-chemical Industry, 1880-1910: Studies in the American Technological Environment.* Westport: Greenwood Press, 1981. xxxviii + 390 pp. Appendices, Bibliography, Index.

Presents a study of the context of industrial change rather than a study of an industry. Argues that the electrochemical industry may have been the first industry to be denoted by a process rather than a product. Credits the American electrochemical industry with the leadership role in the adoption of the concept of unit operations--one of the major distinguishing features of modern chemical manufacturing--into the chemical industry. Suffers from problems of organization.

516. Wilkinson, Norman B. *Lammot du Pont and the American Explosives Industry, 1850-1884.* Charlottesville: University Press of Virginia, 1984. xii + 332 pp. Index.

Combines biographical detail with an exposition of the development of the Du Pont Company. Discusses both the technological and managerial contributions of Lammot du Pont.

COMMUNICATIONS

517. Aitken, Hugh G. J. *The Continuous Wave: Technology and American Radio, 1900-1932.* Princeton: Princeton University Press, 1985. xvii + 588 pp. Appendix, Index.

Integrates technical detail and biographical material in tracing the history of the transformation of radio technology from the spark system to the continuous wave. Finds science's role in the transformation was very small. Provides an account of the early years of the Radio Corporation of America. Views the radio as a case study of the management of technology. This is the standard history to date of this technology.

518. Czitrom, Daniel J. *Media and the American Mind: From Morse to McLuhan.* Chapel Hill: University of North Carolina Press, 1982. xiv + 254 pp. Bibliography, Index.

Analyzes the contemporary American response to the electric telegraph, motion pictures, and radio. Identifies three major traditions in American thought regarding communication: communication as an agent for reform and restoring consensus at the turn of the century, the empirical, social science approach prevalent from the 1930s through 1960, and the influence of Harold Innis and Marshall McLuhan.

519. Douglas, Susan J. "Amateur Operators and American Broadcasting: Shaping the Future of Radio." *Imagining Tomorrow: History, Technology, and the American Future* (item 049), pp. 35-57.

Discusses the predictions that radio communication would reshape life by transcending isolation, both physical and otherwise. Finds that amateur radio operators did utilize the radio in some of the ways predicted by the press.

520. Douglas, Susan J. "Technological Innovation and Organizational Change: The Navy's Adoption of Radio, 1899-1919." *Military Enterprise and Technological Change: Perspectives on the American Experience* (item 176), pp. 117-173.

Describes a situation where a bureaucracy confronts a new technology which will alter its traditional command and control arrangements.

521. Frommer, Myrna. "How Well Do Inventors Understand the Cultural Consequences of Their Inventions? A Study of: Samuel Finley Breese Morse and the Telegraph, Thomas Alva Edison and the Phonograph, and Alexander Graham Bell and the Telephone." Ph.D. dissertation, New York University, 1987.

Finds that in all three cases, the applications of the inventions were very different from those anticipated by the inventors. Concludes that inventors have no control over the fate of their inventions.

522. Joel, A. E., Jr. *A History of Engineering and Science in the Bell System.* Volume 3: *Switching Technology (1925-1975).* Whippany, N.J.: Bell Telephone Laboratories, 1982. xiv+639 pp. Chronology, Index.

Examines the evolution of automatic telephone switching technology from electromechanical dial systems to electronic networks. Assumes background knowledge and technical competence on the part of the reader. This is a continuation of Item I:718.

523. Kielbowicz, Richard B. "News Gathering by Mail in the Age of the Telegraph: Adapting to a New Technology." *Technology and Culture,* 28 (1987): 26-41.

Adopts niche theory from ecology to explain the response of the mails to the challenge of the telegraph. Argues that the mails and the telegraph had geographically and functionally differentiated niches. The telegraph carried the news to newspapers at the terminuses of the lines, then the newspapers would be distributed through the mails to more distant points. The mails handled news which was more complex, opinionated, and of narrower interest than that carried by the telegraph.

524. McGinn, Robert E. "Stokowski and the Bell Telephone Laboratories: Collaboration in the Development of High-Fidelity Sound Reproduction." *Technology and Culture,* 24 (1983): 38-75.

Views the collaboration of Leopold Stokowski and Bell Laboratories as part of Bell's transformation of audio technology from craft-based cut-and-try to a profession based on science and systems engineering. Sees this incident as helping legitimate the artist-scientist/engineer collaboration.

525. Marvin, Carolyn. "Dazzling the Multitude: Imaging the Electric Light as a Communications Medium." *Imagining Tomorrow: History, Technology, and the American Future* (item 049), pp. 202-217.

Discusses the predictions in the late nineteenth and early twentieth centuries that the electric light would be a form of communication.

526. Udelson, Joseph H. *The Great Television Race: A History of the American Television Industry, 1925-1941.* University: University of Alabama Press, 1982. xiii+197 pp. Bibliography, Index.

Traces the history of the technology until the point where the major problems were solved and the Federal Communications Commission approved commercial television. Focuses on the engineering and manufacturing aspects of television. Illustrates the role of political consensus in gaining approval.

527. Wasserman, Neil H. *From Invention to Innovation: Long-Distance Telephone Transmission at the Turn of the Century.* Baltimore: Johns Hopkins University Press, 1985. xxiii + 160 pp. Appendices, Index.

Argues that the experience with long-distance transmission demonstrated the need for a higher level of technical sophistication at AT&T. Uses the loading coil invention to test his model for the process of invention and innovation in a modern corporation.

COMPUTERS

528. Ceruzzi, Paul. "An Unforeseen Revolution: Computers and Expectations, 1935-1985." *Imagining Tomorrow: History, Technology, and the American Future* (item 049), pp. 188-201.

Contends that the inventors of the computer saw it as a piece of apparatus for a specific purpose, failing to recognize potential applications.

529. Owens, Larry. "Vannevar Bush and the Differential Analyzer: The Text and Context of an Early Computer." *Technology and Culture,* 27 (1986): 63-95.

Characterizes the Rockefeller Differential Analyzer as a culmination of Bush's work rather than an aborted beginning. Describes it as an embodiment of an engineering culture.

530. Pugh, Emerson W. *Memories that Shaped an Industry: Decisions Leading to the IBM System /360.* Cambridge: MIT Press, 1984. x + 323 pp. Chronology, Index.

Wishes to document the invention of the memory core, the key to the stored-program computer. Relies heavily on internal IBM documents. Uses the memory development to demonstrate the successful management system of IBM.

531. Redmond, Kent C., and Thomas M. Smith. *Project Whirlwind: The History of a Pioneer Computer.* Bedford, Mass.: Digital Press, 1980. vii + 280 pp. Appendices, Index.

Focuses on the administrative history of the first high-speed digital computer able to operate in "real time." Looks at the events from the perspective of Jay Forrester and Robert R. Everett, the directors of the project. Relies considerably on oral history.

532. Stern, Nancy. *From ENIAC to UNIVAC: An Appraisal of the Eckert-Mauchly Computers*. Bedford, Mass.: Digital Press, 1981. ix+286 pp. Appendix, Bibliography, Index.

Describes the institutional, technological, scientific, and economic problems facing John Presper Eckert, Jr., and John William Mauchly in developing four electronic digital computers between 1943 and 1951. Provides a running commentary on the story from the 1973 patent case, Honeywell v. Sperry Rand. Argues that computer development was a response to a technological need, not the result of scientific discoveries. Disagrees with the court's decision in the patent case.

533. Wildes, Karl L., and Nilo A. Lindgren. *A Century of Electrical Engineering and Computer Science at MIT, 1882-1982.* Cambridge: MIT Press, 1985. xi+423 pp. Appendix, Index.

Chronicles the history of one of the central institutions in the development of American electrical engineering. Highlights the research contributions of individual faculty.

DOMESTIC TECHNOLOGY

534. Boydston, Jeanne. "Home and Work: The Industrialization of Housework in the Northeastern United States from the Colonial Period to the Civil War." Ph.D. dissertation, Yale University, 1984.

Analyzes the impact of industrialization on the economic function of housework, examining both laboring-class and middle-class households. Looks at both the ideological conceptions of housework and the value of it both within households and to the larger industrial economy.

Published as *Home and Work: Housework, Wages, and the Ideology of Labor in the Early Republic.* New York: Oxford University Press, 1990. xxii+222 pp. Bibliography, Index.

535. Busch, Jane. "Cooking Competition: Technology on the
      Domestic Market in the 1930s." *Technology and Culture,*
      24 (1983): 222-245.

      Finds that competition between gas and electric ranges
      led to innovation and convergence. Shows that advertisements
      focused on social status. Concludes that competition resulted in
      better products and more sales.

536. Cooper, Gail Ann. "'Manufactured Weather': A History of Air
      Conditioning in the United States, 1902-1955." Ph.D.
      dissertation, University of California, Santa Barbara,
      1987.

      Focuses on air conditioning as artificial climate. Looks
      at industrial, commercial, and residential usages. Shows how
      the technology evolved to meet changing concepts of acceptable
      levels of comfort.

537. Cowan, Ruth Schwartz. "Ellen Swallow Richards: Technology
      and Women." *Technology in America: A History of
      Individuals and Ideas* (item 073), pp. 142-150.

      Highlights the changes in American technology which
      resulted from Richards's ideas and the education movement--
      home economics--founded on her ideas. Observes that the United
      States was in transition from a servant-based household to one
      in which an educated housewife ran the home using science and
      technology.

538. Cowan, Ruth Schwartz. *More Work for Mother: The Ironies of
      Household Technology from the Open Hearth to the
      Microwave.* New York: Basic Books, 1983. xiv+257 pp.
      Bibliography, Index.

      Traces the history of the industrialization of housework
      from 1860. Argues that labor-saving devices often reorganized
      work, shifted some responsibilities from men to women, replaced
      servants, and eliminated drudgery, but did not reduce the total
      amount of household labor for women. Technology provided a
      higher standard of living rather than less work.

539. Horrigan, Brian. "The Home of Tomorrow, 1927-1945."
      *Imagining Tomorrow: History, Technology, and the
      American Future* (item 049), pp. 137-163.

Focuses on a period of a shortage of housing and revolution in design. Uses Buckminster Fuller's Dymaxion House as a starting point.

540. Hoy, Suellen. "The Garbage Disposer, the Public Health, and the Good Life." *Technology and Culture*, 26 (1985): 758-784.

Sees the disposer's appeal partly as a solution to the problem of urban garbage, partly as a laborsaving device making middle-class life easier at a time when more middle-class women were working.

541. McGaw, Judith A. "Women and the History of American Technology." *Signs: Journal of Women in Culture and Society*, 7 (1982): 798-828.

Essay review. Concentrates on the technology of homemaking and the technology of the non-domestic workplace. Focuses on the literature published after 1969. Complains that much of the work either discusses technology without its social context or technology-based social change without considering the technology itself. Evaluates the literature and identifies needs.

542. Rose, Mark H. "Urban Environments and Technological Innovation: Energy Choices in Denver and Kansas City, 1900-1940." *Technology and Culture*, 25 (1984): 503-539.

Argues that the availability of gas and electric service was crucial to the construction of the modern domestic environment in these two cities. Claims innovation--both technical and organizational--by the utilities was in response to the social structure and geography of the cities. Finds that cities had zones defined by income and utility service.

543. Schroeder, Fred E. H. "More 'Small Things Forgotten': Domestic Electrical Plugs and Receptacles, 1881-1931." *Technology and Culture*, 27 (1986): 525-543.

Traces the evolution of electrical plugs from permanently wired electrical appliances to standardized, interchangeable plugs and outlets. Argues that the standardization of plugs and receptacles was a necessary step in the development of the modern, technologically dependent household.

544. Strasser, Susan. "An Enlarged Human Existence? Technology and Household Work in Nineteenth-Century America." *Women and Household Labor.* Edited by Sarah Fenstermaker Berk. Beverly Hills: Sage, 1980, pp. 29-51.

Rejects claims that the available technology was widely diffused. Finds that the technology had little impact on women's work except in the case of the wealthy.

545. Thrall, Charles A. "The Conservative Use of Modern Household Technology." *Technology and Culture,* 23 (1982): 175-194.

Utilizes a late 1960s survey of households. Finds that the development of household technology has had a conservative effect by making it easier for those who are stereotyped for certain activities to do them without any assistance.

546. Upton, Dell. "Traditional Timber Framing." *Material Culture of the Wooden Age* (item 056), pp. 35-93.

Places framing practices within cultural milieus. Argues that the introduction of balloon framing and mass-produced building materials did not represent so much the intrusion of a radical system into folk carpentry, as a new stage in a continuing effort to find the best structural system.

ELECTRICAL TECHNOLOGY

547. Belfield, Robert Blake. "The Niagara Frontier: The Evolution of Electric Power Systems in New York and Ontario, 1880-1935." Ph.D. dissertation, University of Pennsylvania, 1981.

Observes that during the beginning of the period, American technology, in the form of the universal electric power system, was transferred to Canada. Finds that by the end of the period the Ontario Hydro's model of public power and rural and domestic electrification programs had been transferred south in the form of the Power Authority of the State of New York and, ultimately, the Tennessee Valley Authority.

548. Christie, Jean. "Morris L. Cooke and Energy for America." *Technology in America: A History of Individuals and Ideas* (item 073), pp. 202-212.

Highlights the activities of "Power Progressives" (reformers concerned with applying science to human needs), which culminated in the establishment of the Tennessee Valley Authority and Rural Electrification Administration.

549. Friedel, Robert, and Paul Israel. *Edison's Electric Light: Biography of an Invention.* New Brunswick: Rutgers University Press, 1986. xvi + 263 pp. Bibliography, Index.

Views Edison as the transition between the heroic inventor and the research and development laboratory. Provides a detailed account of the invention of the electric light system based on the contemporary archival record, especially the pictorial record. Demonstrates the significance of nonverbal documentation in the understanding of the history of technology.

550. Hirsh, Richard F. "Conserving Kilowatts: The Electric Power Industry in Transition." *Energy in American History* (item 050), pp. 295-305.

Argues that technological stasis--a phase when limits to progress are reached--occurred in the electric power industry during the late 1960s.

551. Hughes, Thomas P. *Networks of Power: Electrification in Western Society, 1880-1930.* Baltimore: Johns Hopkins University Press, 1983. xi + 474 pp. Index.

Incorporates the author's previous research on technological systems, reverse salients, technological momentum, and regional style. Presents system formation and growth as a four-stage process: invention and development, technology transfer, system growth, and system momentum. Contrasts the successful transplantation of Edison's electrification system to Berlin with the failure in London. Blames the different results on political rather than technological or economic causes. Examines the social, economic, and political forces that shape electric power systems in the United States (Chicago), Germany (Berlin) and England (London). Finds that the configuration of the system was determined by the relative strengths of political authorities and the utility managers.

552. McMahon, A. Michal. *The Making of a Profession: A Century of Electrical Engineering in America.* New York: Institute of Electrical and Electronics Engineers Press, 1984. xv + 304 pp. Index.

Provides a centennial history of the Institute of Electrical and Electronics Engineers. Divides the history into four eras: telegraphy, power engineering, radio, and electronics. Examines how the values of the electrical engineering profession changed over time, focusing on how non-engineers (businessmen or the military) managed to establish professional agendas. Concentrates on leading members of the Institute of Electrical and Electronics Engineers, not the rank-and-file.

553. Schallenberg, Richard H. "The Anomalous Storage Battery: An American Lag in Early Electrical Engineering." *Technology and Culture,* 22 (1981): 725-752.

Analyzes the only area of electrical technology in which the United States lagged behind Europe in the 1880s and early 1890s--the manufacture and use of storage batteries. Blames the lag on Thomas Edison's opposition to using batteries for auxiliary power in lighting systems, which was the European practice, because he felt they were inherently uneconomical. Credits the transformation of the storage battery industry in the United States to the use of batteries in electric trolleys.

ENGINEERING: GENERAL, MECHANICAL, AND CIVIL

554. Christie, Jean. *Morris Llewellyn Cooke: A Progressive Engineer.* New York: Garland, 1983. 272 pp. Bibliography, Index.

Presents Cooke as a promoter of national planning and an advocate of the function of the engineer as solver of the problems of industrial societies. Describes his objectives as the mitigation of the evil consequences of capitalism and the increase in the distribution of the benefits of the system. Discusses his advocacy of the role of the engineer in the conservation of natural resources.

555. Kemp, Emory L. "Roebling, Ellet, and the Wire-Suspension Bridge." *Bridge to the Future: A Centennial Celebration of the Brooklyn Bridge* (item 558), pp. 41-62.

Traces the events which led to the building of long-span, wire-suspension bridges in the United States. Discerns a French influence.

556. Kouwenhoven, John A. "The Designing of the Eads Bridge." *Technology and Culture,* 23 (1982): 535-568.

Concludes that James Buchanan Eads did design the 1874 bridge across the Mississippi at St. Louis.

557. Kouwenhoven, John A. "James Buchanan Eads: The Engineer as Entrepreneur." *Technology in America: A History of Individuals and Ideas* (item 073), pp. 80-91.

Argues that understanding Eads's career depends upon the realization that he combined technical ability with financial genius.

558. Latimer, Margaret, Brooke Hindle, and Melvin Kranzberg, editors. *Bridge to the Future: A Centennial Celebration of the Brooklyn Bridge.* New York: New York Academy of Sciences, 1984 (Annals of the New York Academy of Sciences, 424). xiv+355 pp. Index.

Contains twenty-one articles focusing on the building of the bridge, the development of new technology, the role of science and design in engineering, technological symbols, and the impact of transportation on society. Includes items 093, 555, 568.

559. Layton, Edwin T., Jr. "European Origins of the American Engineering Style of the Nineteenth Century." *Scientific Colonialism: A Cross-Cultural Comparison.* Edited by Nathan Reingold and Marc Rothenberg. Washington, D.C.: Smithsonian Institution Press, 1987, pp. 151-166.

Hypothesizes that one of the primary causes of national differences in technology is the adapting of technology to provide a better fit for a particular society or to minimize harmful effects. Provides an example of American engineers assimilating Continental traditions of mathematical analysis within the American anti-theoretical tradition.

560. Levy, Richard Michael. "The Professionalization of American Architects and Civil Engineers, 1865-1917." Ph.D. dissertation, University of California, Berkeley, 1980.

Looks at two groups within the construction industry. Argues that for the leaders of the American Institute of Architects, architecture was a fine art; they emphasized the aesthetics of design. Contrasts this with the American Society of Civil Engineers, who emphasized sound economic judgement and appropriate use of technology.

561. McMath, Robert C., Jr., et al. *Engineering the New South: Georgia Tech, 1885-1985.* Athens: University of Georgia Press, 1985. xii+560 pp. Bibliography, Index.

   Organizes the discussion by presidential administrations. Considers a wide range of issues, including finances, integration, and the role of sports. Sees Georgia Tech as part of the "New South Creed."

562. Meier, Michael Thomas. "Caleb Goldsmith Forshey: Engineer of the Old Southwest, 1813-1881." Ph.D. dissertation, Memphis State University, 1982.

   Relates the life of a civil engineer and student of hydraulics. Argues that natural science and civil engineering in the American Southwest were closely related. Sees Forshey as a member of that part of the engineering community which supported government sponsorship of internal improvements.

563. Pershey, Edward Jay. "The Early Telescope Work of Warner and Swasey." Ph.D. dissertation, Case Western Reserve University, 1982.

   Investigates the professionalization of mechanical engineering through a study of the machine-tool firm of Worcester Warner and Ambrose Swasey. Characterizes the partners as machinists trained through apprenticeship who adopted many of the values and methods of the college-trained engineers. Credits them with developing a distinctive style of telescopes.

564. Pursell, Carroll W., Jr. "'What the Senate Is to the American Commonwealth': A National Academy of Engineers." *New Perspectives on Technology and American Culture* (item 075), pp. 19-29.

   Discusses the futile campaign in 1917 to establish an engineering equivalent of the National Academy of Sciences. Finds that the concept was attacked both by those who saw no need to reform engineering and by the reformers who saw the suggested academy as undemocratic.

565. Sinclair, Bruce. "Inventing a Genteel Tradition: MIT Crosses the River." *New Perspectives on Technology and American Culture* (item 075), pp. 1-18.

Views the founding and subsequent relocation of the Massachusetts Institute of Technology in the context of the frustrated efforts of engineers to obtain the status of members of high culture, similar to scientists.

566. Sinclair, Bruce. "Local History and National Culture: Notions in Engineering Professionalism in America." *Technology and Culture*, 27 (1986): 683-693.

Points out that we know little about the rank and file of American engineers. Suggests studying local engineering associations.

567. Sokal, Michael M. "Companions in Zealous Research, 1886-1986." *American Scientist*, 74 (1986): 486-508.

Traces the history of Sigma Xi from its founding as an attempt to bolster the status of college-educated engineers in their competition with shop-trained engineers through the continuing controversy over the role of the society in the scientific community. Credits Henry Shaler Williams with shaping the early society. Discusses the clashes between the local chapters and the central administration.

568. Vogel, Robert M. "Designing Brooklyn Bridge." *Bridge to the Future: A Centennial Celebration of the Brooklyn Bridge* (item 558), pp. 3-39.

Describes the evolution of the design of the subaqueous foundations of the towers and the suspended superstructure. Sees the Brooklyn Bridge as an exemplar of the belief that the best design is the simplest.

INDUSTRIAL RESEARCH

569. Dennis, Michael Aaron. "Accounting for Research: New Histories of Corporate Laboratories and the Social History of American Science." *Social Studies of Science*, 17 (1987): 479-518.

Critiques the literature dealing with science and industry prior to World War I. Argues that difficulties arise in those discussions because of their use of Robert Merton's sociology of science for an inappropriate institutional structure. Suggests that fuller understanding of the history of corporate laboratories would result from looking at the broader issue of research activities separate from education.

570. Hoddeson, Lillian H. "The Discovery of the Point-Contact Transistor." *Historical Studies in the Physical Sciences,* 1981 (12): 41-76.

   Reviews the scientific background, institutional setting, and research steps that led to the discovery of the point-contact transistor by John Bardeen and Walter Brattain. Characterizes the invention as the first meaningful accomplishment of the reorganized Physical Research Department of the Bell Telephone Laboratories. Describes the two most important features of the department as an orientation towards basic research in solid-state physics and an emphasis on multidisciplinary team research.

571. Hoddeson, Lillian H. "The Emergence of Basic Research in the Bell Telephone System, 1875-1925." *Technology and Culture,* 22 (1981): 512-544.

   Argues that the research program at Bell evolved gradually in response to particular technical telephone problems. Maintains that commercially defined objectives and social developments gave urgency and relevance to the technical problems which the research solved.

572. Reich, Leonard S. *The Making of American Industrial Research: Science and Business at GE and Bell, 1876-1926.* Cambridge: Cambridge University Press, 1985. xvi + 309 pp. Index.

   Argues that industrial research was an integral part of corporate strategy. Contrasts General Electric's pluralistic style of laboratory management to Bell Laboratories's tight control and narrow focus. Links the contrasting styles to the contrasting needs of the two corporations and demonstrates that corporate structure and objectives influence direction and organization of research. Sees the ultimate use of research in both corporations to maintain quasi-monopolistic positions, not advance science or provide for the advantage in a competitive market.

573. Russo, Arturo. "Fundamental Research at Bell Laboratories: The Discovery of Electron Diffraction." *Historical Studies in the Physical Sciences,* 12 (1981): 117-160

   Compares and contrasts the simultaneous discovery of electron diffraction in 1927 by two American physicists working at Bell Laboratories and an English physicist in Scotland. Argues that the English physicist George P. Thomson followed a linear path from theory to experiment, but the American discovery was by chance.

574. Smith, John K. "The Ten-Year Invention: Neoprene and Du Pont Research, 1930-1939." *Technology and Culture,* 26 (1985): 34-55.

Traces the transformation of an idea into a profitable product. Credits the success of neoprene to control of the product and the use of a coherent commercialization strategy.

575. Vincenti, Walter G. "The Davis Wing and the Problem of Airfoil Design: Uncertainty and Growth in Engineering Knowledge." *Technology and Culture,* 27 (1986): 717-758.

Looks at the design process at Consolidated Aircraft Corporation and the work of David R. Davis in the 1930s. Finds that Davis's work neither relied on theoretical fluid mechanics nor experimental data. Concludes that the engineers at Consolidated responded to the uncertainties in knowledge by relying on both test data and non-engineering factors.

576. Wise, George. *Willis R. Whitney, General Electric, and the Origins of U.S. Industrial Research.* New York: Columbia University Press, 1985. xi+375 pp. Appendices, Bibliography, Index.

Argues that Whitney failed to develop an academic career because he was too much of a generalist. Sees the novelty of General Electric Laboratory as an interest in the discovery of new principles rather than only in the application of new principles. Describes Whitney as a research director who served as a bridge between pure and applied science, blending aspects of the corporation and the academic laboratory. Denies that the research laboratory of the 1920s was either a stop on the technological assembly line or a university in exile. Presents Whitney as using his work as a substitute for an unhappy personal life.

INDUSTRIALIZATION

577. Chandler, Alfred D. Jr. "Technology and the Transformation of Industrial Organization." *Technology, the Economy, and Society: The American Experience* (item 047), pp. 56-82.

Presents the emergence of the modern corporation and the oligopolistic structure within certain industries as a response to innovations in the technologies of mass production and distribution. Dates this process to the late nineteenth century.

578. Cochran, Thomas C. "Cotton Textiles and Industrialism." *Science and Society in Early America: Essays in Honor of Whitfield J. Bell, Jr.* (item 025), pp. 251-269.

Rejects the thesis that American has an industrial revolution in favor of the argument that there was continuous growth. Blames the incorrect historiography which emphasizes the New England cotton textile industry on the historical consciousness of New England, which resulted in the creation and preservation of records.

579. Cochran, Thomas C. *Frontiers of Change: Early Industrialism in America.* New York: Oxford University Press, 1981. vii + 179 pp. Bibliography, Index.

Argues that the United States met and surpassed Europe during the antebellum period because of a combination of favorable geography. natural resources, and culture (Americans were open to innovation and risks, even among the elite). Expresses skepticism with the contention that the shortage of skilled labor was the driving motivation to industrialization.

580. Ferguson, Eugene S. *Oliver Evans, Innovative Genius of the American Industrial Revolution.* Wilmington: Hagley Museum, 1980. 72 pp. Bibliography.

Summarizes his life and contributions. Sees Evans as trying to go beyond the patent system by searching for an institution which would ensure continuing technological innovation in the American context.

581. Glasgow, Jon. "Innovation on the Frontier of the American Manufacturing Belt." *Pennsylvania History,* 52 (1985): 1-21.

Proposes an Industrial Frontier Hypothesis. Theorizes that the demand for raw materials by industries in the core spurred a growth of primary industries in the adjacent periphery. Presents case studies, including the automobile industry, anthracite iron, and Bessemer steel.

582. Gordon, Robert B. "Cost and Use of Water Power during Industrialization in New England and Great Britain: A Geological Interpretation." *Economic History Review*, 36 (1983): 240-259.

Concludes that the physical limitations on the availability of water power at low cost was not a problem. Believes that steam power replaced water power because it provided convenience.

583. Gordon, Robert B. "Hydrological Science and the Development of Waterpower for Manufacturing." *Technology and Culture*, 26 (1985): 204-235.

Analyzes three case studies of New England manufacturers. Concludes that in all these cases the available waterpower was not fully developed because the knowledge of hydrology and geology was inadequate for that purpose.

584. Greenberg, Dolores. "Reassessing the Power Patterns of the Industrial Revolution: An Anglo-American Comparison." *American Historical Review*, 87 (1982): 1237-1261.

Finds more dependency on renewable energy sources during the Industrial Revolution than other scholars have indicated. Argues that animate power continued to be an important source of power throughout the nineteenth century. Challenges the belief that steam power was vital for the Industrial Revolution. Believes the expanded use of traditional power sources was characteristic of the period.

585. Hindle, Brooke. "The American Industrial Revolution Through Its Survivals." *Science and Society in Early America: Essays in Honor of Whitfield J. Bell, Jr.* (item 025), pp. 271-310.

Argues for the importance of material culture in historical research. Presents a precis of the exhibition "Engines of Change" (see item 586).

586. Hindle, Brooke, and Steven Lubar. *Engines of Change: The American Industrial Revolution.* Washington, D.C.: Smithsonian Institution Press, 1986. 309 pp. Bibliography, Index.

Surveys, summarizes, and popularizes current scholarship. Attempts to provide context for the three-dimensional artifacts of the American Industrial Revolution. Relies heavily on illustrations.

587. Hunter, Louis C. *A History of Industrial Power in the United States, 1780-1930.* Volume II: *Steam Power.* Charlottesville: The University Press of Virginia. xxiii+732 pp. Appendices, Index.

Examines the history of the stationary steam engine in the United States from approximately 1750 to 1900. Analyzes in detail the building of steam engines, the contributions of George H. Corliss, the furnace boiler, fuels, and the impact of steam upon urban water supply. Compares and contrasts the use of steam in Great Britain and the United States. Includes a great deal of statistical information.

588. Jeremy, David J. *Transatlantic Industrial Revolution: The Diffusion of Textile Technologies between Britain and America, 1790-1830s.* Cambridge: The MIT Press, 1981. xvii+384 pp. Appendices, Glossary, Bibliography, Index.

Documents the factors hampering or encouraging the flow of textile technology. Identifies four stages in the transfer of technology: introduction, trial factories, widespread diffusion, and the modification of the technology to fit local conditions. Emphasizes the role of individual immigrants.

589. Kulik, Gary. "A Factory System of Wood: Cultural and Technological Change in the Building of the First Cotton Mills." *Material Culture of the Wooden Age* (item 056), pp. 300-335.

Demonstrates the omnipresence of wood in American factories during early industrialization. Argues that one characteristic of American technology was its resource intensity; in this case, wood. Contends that the architecture of early factories was chosen to downplay the revolutionary changes being made to life.

590. Layton, Edwin. "The Industrial Evolution in America: Energy in the Age of Water and Wood." *Energy in American History* (item 050), pp. 249-263.

Divides the pre-1870 history of American energy use into three periods: an age of low energy density or "frontier technology," until 1750; the "Pre-Industrial Revolution" period, 1750-1789; and the years between 1789 and 1870, when the industry in the United States was fueled by water and wood. Notes that coal did not replace water as the most important source of industrial power until about 1870.

591. Lubar, Steve. "Culture and Technological Design in the 19th-century Pin Industry: John Howe and the Howe Manufacturing Company." *Technology and Culture,* 28 (1987): 253-282.

Explores the influence of the larger culture on the technological style (design of machines) and industrial style (the interaction of management, workers, and technology) of pin manufacturing. Chooses that industry because there were numerous approaches possible. Concludes that Howe's machines were influenced by or reflected the social structure of the industry, the skills and experience of mechanicians, and familiar and widespread methods of solving mechanical problems. Sees the industrial style influenced by larger trends towards centralized control and use of less skilled workers.

592. McGaw, Judith A. *Most Wonderful Machine: Mechanization and Social Change in Berkshire Paper Making, 1801-1885.* Princeton: Princeton University Press, 1987. xvii+439 pp. Appendices, Bibliography, Index.

Views the mechanization of the paper industry and its consequences as representative of mechanization in the United States as a whole. Relates success in the mechanized business to membership in networks of owners. Argues that mechanization occurred because of abundant natural resources, limited skilled workers, and widespread machine-making skills. Finds no evidence of class-consciousness arising among the workers, although mill workers had less hope of becoming owners and were in more danger of injury as mechanization proceeded. Finds that women's contributions to the paper-making industry were essential to the mechanization, but were discounted by the male community. Concludes that the social consequences of mechanization were abetted by technological change, not determined by it. Argues for the role of human choice and traditional cultural values in shaping the mechanized work place.

593. Scranton, Philip. *Proprietary Capitalism: The Textile Manufacture at Philadelphia, 1800-1885.* Cambridge: Cambridge University Press, 1983. xiii+431 pp. Index.

Introduces the concept of "accumulation matrix," which is the total of the social and economic factors which constitute the total situation facing businessmen. Includes material elements, sociocultural context, and external events; some of these factors are controllable by businessmen, some are not. Contrasts the textile mills in Philadelphia, which were small with lots of skilled workers and relatively little capital outlay per worker, with the larger mills, high capital outlay and unskilled labor force of Lowell. Finds that the Philadelphia textile industry was characterized by flexibility and specialization.

594. Tucker, Barbara M. *Samuel Slater and the Origins of the American Textile Industry, 1790-1860.* Ithaca: Cornell University Press, 1984. 268 pp. Index.

Argues that the relationship between labor and management in Slater's factory system was characterized by the persistence of tradition--with the integration of preindustrial institutions and customs into the new industrial order--and stability. Identifies among the persistent traditions the influence of the kinship unit over work, including allocation of jobs and discipline. Ascribes the changes in the factory system in the 1830s, after Slater's death, to his sons' dismantling of the patriarchal order for a more rational economic system. Claims that flight was the most persistent form of protest to the dissolution of the traditional prerogatives.

INSTRUMENT MAKERS

595. Bedini, Silvio A. *At the Sign of the Compass and Quadrant: The Life and Times of Anthony Lamb.* Philadelphia: American Philosophical Society, 1984 (Transactions of the American Philosophical Society, 74, Part 1). 84 pp. Index.

Traces the life and background of the only professionally trained mathematical instrument maker in the American colonies. Provides considerable information on the colonial instrument trade.

596. Warner, Deborah Jean. "Optics in Philadelphia During the Nineteenth Century." *Proceedings of the American Philosophical Society,* 129 (1985): 291-299.

Concludes that the collaboration by artisans, engineers, scientists, businessmen and government necessary for the development of a precision optical industry never occurred.

597. Warner, Deborah Jean. "Rowland's Gratings: Contemporary Technology." *Vistas in Astronomy,* 1986, 29:125-130.

Describes Henry A. Rowland's predecessors. Calls Rowland's concave diffraction grating the first American-made precision scientific instrument that was superior to its European counterparts.

IRON, STEEL, AND METALLURGY

598. Gordon, Robert B. "Material for Manufacturing: The Response of the Connecticut Iron Industry to Technological Change and Limited Resources." *Technology and Culture,* 24 (1983): 602-634.

Concludes that the iron industry of the Salisbury district survived, despite limited resources, because it was able to demand premium prices for its product due to its superior reputation. Finds that the industry closed down when metallurgical advances led to the questioning of that superiority. Utilizes samples of Salisbury district iron production as data.

599. Loveday, Amos J., Jr. *The Rise and Decline of the American Cut Nail Industry: A Study of the Interrelationships of Technology, Business Organization, and Management Techniques.* Westport, Conn.: Greenwood Press, 1983. xx + 160 pp. Bibliography, Index.

Provides a micro-history of the industry through World War I. Focuses on the industry around Wheeling, West Virginia. Integrates discussions of changes in technology, markets, management and accounting techniques, and labor organization.

600. Misa, Thomas Jay. "Science, Technology and Industrial Structure: Steelmaking in America, 1870-1925." Ph.D. dissertation, University of Pennsylvania, 1987.

Focuses on the qualitative changes in the science, technology, and industrial structure of steelmaking. Looks at five users of five different segments of the steel industry: railroads (Bessemer steel), the navy (open hearth and case hardened steel), machine builders (crucible steel), structural trade (ductile steel with low phosphorus), and automobiles (large quantities of high-quality steel produced by electric furnaces). Argues that after 1900 steelmakers embraced business strategies grounded on the quality of the product, rather than only quantity and cost. Contends that these strategies led to the adoption of the new technological processes and the expansion of the role of science in the industry.

601. Mulholland, James A. *A History of Metals in Colonial America.* University: University of Alabama Press, 1981. xvi+216 pp. Bibliography.

Contends that metals served as an essential ingredient in the material culture of western civilization. Emphasizes gold, silver, and iron. Sees military power as an accompaniment to the search for precious metals. Traces the changing approach to iron in the colonies from exporting ore, to the export of pig iron, and finally to the domestic use of iron.

602. Tweedale, Geoffrey. *Sheffield Steel and America: A Century of Commercial and Technological Interdependence, 1830-1930.* New York: Cambridge University Press, 1987. xvi+296 pp. Appendix, Bibliography, Index.

Focuses on the important role of specialty steel (crucible steel or alloy steel) on the American economy. Argues that American efforts to copy the products of Sheffield, England, were unsuccessful until the last third of the nineteenth century and that dependence on the English continued into the twentieth century. Contrasts the Sheffield situation--the resistance of the workers to allow mechanization, and the failure of the businessmen to integrate their concerns into larger organizations and to advertise--with the post-1880, mechanized, integrated, heavily advertised American industry. Blames the ultimate decline of Sheffield's position on the invention of the high-frequency electric furnace.

MASS PRODUCTION AND THE AMERICAN SYSTEM

603. Chandler, Alfred D., Jr. "The American System and Modern Management." *Yankee Enterprise: The Rise of the American System of Manufactures* (item 609), pp. 153-170.

Claims that the complex nature of both the process and products of manufacturing led to the development of modern factory and corporate management techniques in the late nineteenth and early twentieth centuries.

604. Ferguson, Eugene S. "History and Historiography." *Yankee Enterprise: The Rise of the American System of Manufactures* (item 609), pp. 1-23.

Divides the historiography into five topics: interchangeable parts, economic and social explanations, the "New Factory
System," the export of American products and practices to the
rest of the world, and studies of the social costs and benefits.
Provides an excellent summary of the state of the literature in the
late 1970s.

605. Hoke, Donald Robert. "Ingenious Yankees: The Rise of the
    American System of Manufactures in the Private Sector."
    Ph.D. dissertation, University of Wisconsin-Madison,
    1984.

Examines four industries: wooden movement clocks, axes,
watches, and typewriters. Emphasizes the role of mechanics in
the rise of the American System of Manufactures. Argues that
the contributions of the private sector to the development of the
American System have been overlooked in favor of the role of
government armories.

Published as *Ingenious Yankees: The Rise of the American
System of Manufactures in the Private Sector.* xiv+345 pp.
Bibliography, Index. New York: Columbia University Press,
1990.

606. Holley, I. B., Jr. "A Detroit Dream of Mass-produced Fighter
    Aircraft: The XP-75 Fiasco." *Technology and Culture,*
    28 (1987): 578-593.

Describes the failure of General Motors to mass-produce
aircraft from ready-made parts during World War II. Blames the
failure on the differences between aircraft and automobile
manufacturing and the difficulties of simultaneous development
and production.

607. Hounshell, David A. "Ford Eagle Boats and Mass Production
    during World War I." *Military Enterprise and Technologi-
    cal Change: Perspectives on the American Experience*
    (item 176), pp. 175-202.

Blames Ford's failure to mass-produce boats on his failure
to recognize the limits of assembly-line methods and the
differences between car and marine engineering.

608. Hounshell, David A. *From the American System to Mass Production, 1800-1932: The Development of Manufacturing Technology in the United States.* Baltimore: The Johns Hopkins University Press, 1984. xxi+411 pp. Bibliography, Index.

Argues that mass production based on interchangeable parts came much slower and later than previously thought. Discusses manufacturing technology in a number of industries: sewing machines, furniture, reapers, bicycles, and automobiles. Describes the technical aspects of these industries. Relies heavily, and utilizes skillfully, illustrations and other pictorial material.

609. Mayr, Otto, and Robert C. Post, editors. *Yankee Enterprise: The Rise of the American System of Manufactures.* Washington, D.C.: Smithsonian Institution Press, 1981. xx+236 pp. Bibliography, Index.

Stems from a 1978 symposium. Contains nine papers. Includes items 91, 190, 603, 604, 610, 611, 612, 650.

610. Musson, A. E. "British Origins." *Yankee Enterprise: The Rise of the American System of Manufactures* (item 609), pp. 25-48.

Argues that mass-production engineering originated in Great Britain. Points to American dependence on British technology.

611. Rosenberg, Nathan. "Why in America?" *Yankee Enterprise: The Rise of the American System of Manufactures* (item 609), pp. 49-61.

Views the American System of Manufacturing as part of a larger economic evolution driven by the high demand for certain types of commodities and the abundance of natural resources.

612. Uselding, Paul. "Measuring Techniques and Manufacturing Practices." *Yankee Enterprise: The Rise of the American System of Manufactures* (item 609), pp. 103-126.

Argues that precision machinery was dependent upon the prior existence of accurate measurement devices and methods. Contrasts the English making system, where measurement was used to make parts fit each other, with the American manufacturing system, in which parts were made to specific dimensions.

Compares the line measurement techniques of the making system with the end measurement techniques of the manufacturing system.

NUCLEAR ENERGY

613. Boyer, Paul S. *By the Bomb's Early Light: American Thought and Culture at the Dawn of the Atomic Age.* New York: Pantheon, 1985. xx+440 pp. Index.

Focuses on the period 1945-1950. Finds that Americans quickly understood the threat the A-Bomb represented to the survival of humanity. Organizes the discussion around responses to the bomb: world government movements, the Federation of Atomic Scientists, searches for beneficial aspects of atomic technology, the social implication of atomic energy, the impact upon morals and values, and its relatively minor influence on American culture (especially literature), and finally, acceptance. Concludes that this was the first of three periods of activism and concern about the threat of atomic/nuclear weapons, each giving way to a period of apathy.

614. Cantelon, Philip L., and Robert C. Williams. *Crisis Contained: The Department of Energy at Three Mile Island.* Carbondale: Southern Illinois University Press, 1982. xxi+213 pp. Appendices, Bibliography, Index.

Provides a highly detailed account, based on interviews and archival material, of the governmental reaction to the technological failure at the Three Mile Island nuclear power plant. Argues that public fears were intensified because of the breakdown in communications. Commends the response of the Department of Energy.

615. Ford, Daniel. *The Cult of the Atom: The Secret Papers of the Atomic Energy Commission.* New York: Simon and Schuster, 1982. Revised edition, 1984. 273 pp. Index.

Utilizes documents obtained from the Atomic Energy Commission through the Freedom of Information Act to explore the issue of nuclear safety cover-ups. Attacks the Atomic Energy Commission for allowing safety to be sacrificed to economic needs.

616. Malmsheimer, Lonna A. "Three Mile Island: Fact, Frame, and Fiction." *American Quarterly,* 38 (1986): 35-52.

     Utilizes 400 interviews to find that Three Mile Island did not change people's opinions about nuclear power.

617. Mazuzan, George T. "'Very Risky Business': A Power Reactor for New York City." *Technology and Culture,* 27 (1986): 262-284.

     Describes the 1962-1964 effort by Consolidated Edison to build a nuclear power reactor in the middle of New York City, which resulted in the first grass roots protest over the locating of a reactor.

618. Mazuzan, George T., and J. Samuel Walker. *Controlling the Atom: The Beginnings of Nuclear Regulation, 1946-1962.* Berkeley: University of California Press, 1985. x+530 pp. Appendices, Bibliography, Index.

     Provides a detailed, comprehensive account of the development of nuclear technology and the administrative organizations to regulate nuclear power. Discusses the conflict between the Atomic Energy Commission and the Joint Congressional Committee on Atomic Energy, focusing on the political aspect of the dispute between the Republican AEC and the Democratic Committee. Highlights the contradiction of the AEC serving as both regulator and promoter of nuclear power.

619. Mazuzan, George T., and J. Samuel Walker. "Developing Nuclear Power in an Age of Energy Abundance, 1946-1962." *Energy in American History* (item 050), pp. 307-319.

     Explains the rapid development of nuclear power in the United States despite an abundance of alternative fuels. Sees it as a result of the intertwining of nuclear power with America's international prestige and leadership.

TRANSPORTATION

620. Barrett, Paul. *The Automobile and Urban Transit: The Formation of Public Policy in Chicago, 1900-1930.* Philadelphia: Temple University Press, 1983. xiii+295 pp. Bibliography, Index.

Credits public policy, not the automobile, with defeating mass transit. Points out that mass transit in Chicago was a regulated private business, expected to provide inexpensive, convenient service for low fares, while still making a profit.

621. Bezilla, Michael. *Electric Traction on the Pennsylvania Railroad, 1895-1968.* University Park: Pennsylvania State University Press, 1980. vii+233 pp. Appendices, Index.

Focuses on decision making and implementation by the Pennsylvania Railroad. Characterizes the railroad as conservative rather than pioneering, but able to recognize when change was necessary. Finds that the railroad made sound business and financial decisions during the electrification process. Analyzes the evolution of locomotive and car design and the compromises made because of technological limitations.

622. Cheape, Charles W. *Moving the Masses: Urban Public Transit in New York, Boston and Philadelphia, 1880-1912.* Cambridge: Harvard University Press, 1980. vii+287 pp. Bibliography, Index.

Discusses the era of conversion from horsecars to electric streetcars and rapid transit. Emphasizes the social and political aspects of change, not the technological. Concludes that local factors, such as topography and the political environment, shaped the local process.

623. Condit, Carl W. *The Port of New York: A History of the Rail and Terminal System from the Beginnings to Pennsylvania Station.* Chicago: University of Chicago Press, 1980. xvii+456 pp. Bibliography, Index.

Treats New York City as a rail terminus. Argues that the vitality of New York was related to the urban circulation of goods and people. Believes the character of the city was shaped by its metropolitan railroad system.

624. Condit, Carl W. *The Port of New York: A History of the Rail and Terminal System from the Grand Central Electrification to the Present.* Chicago: University of Chicago Press, 1981. xii+399 pp. Bibliography, Index.

Focuses on technological and architectural issues. Argues that electrification was essential to the present configuration of New York City. This is a continuation of item 623.

625. Formwalt, Lee W. "Benjamin Henry Latrobe and the Revival
     of the Gallatin Plan of 1808." *Pennsylvania History,* 48
     (1981): 99-128.

     Uncovers Latrobe's effort to obtain congressional approval
     of a national road and canal system in 1810. Challenges the
     traditional identification of national planning of internal improve-
     ments pre-War of 1812 with Albert Gallatin.

626. Foster, Mark S. *From Streetcar to Superhighway: American City
     Planners and Urban Transportation, 1900-1940.* Philadel-
     phia: Temple University Press, 1981. xiv+246 pp.
     Bibliography, Index.

     Shows how city planners viewed suburbs as a way of
     easing urban congestion. Finds that the planners viewed roads
     as cheaper and more democratic than transit systems. Describes
     the opportunities provided by Great Depression era funding for
     the building of roads between suburbs and cities.

627. Jackson, Kenneth T. "The Impact of Technological Change on
     Urban Form." *Technology, the Economy, and Society: The
     American Experience* (item 047), pp. 150-161.

     Describes the erosion of the walking city, first by the
     trolley, then by the automobile and truck.

628. Kemp, Louis Ward. "Aesthetes and Engineers: The Occupational
     Ideology of Highway Design." *Technology and Culture,*
     27 (1986): 759-797.

     Describes the clash between highway engineers and city
     planners from the birth of the Interstate Road System, in 1944,
     until 1965. Credits the disparity between engineering techniques
     and ideologies to the complexity of urban highway design.

629. Lacey, Robert. *Ford: The Men and the Machine.* Boston: Little,
     Brown and Company, 1986. xix+778 pp. Bibliography,
     Index.

     Presents an unsympathetic picture of Henry Ford, Edsel
     Ford, and Henry Ford II. Blames much of the Ford Company's
     problems after the glory days of the Model T to an obsolete
     management structure.

630. Leslie, Stuart W. *Boss Kettering: Wizard of General Motors.*
New York: Columbia University Press, 1983. xii+382
pp. Index.

Provides an authoritative and scholarly evaluation of
Charles F. Kettering, one of the most significant American
inventors/engineers of the first half of the twentieth century and
the head of the General Motors Research Corporation. Demonstrates Kettering's belief in the importance of small refinements
which have a cumulative effect, the promotion of products, and
attention to markets.

631. Lewis, David L., editor. "The Automobile and American
Culture." *Michigan Quarterly Review,* 19-20 (1980-81):
434-781.

Contains a mixture of scholarly and popular approaches
to the role of the automobile in American culture. Includes
history, fiction, poetry, and autobiography. Addresses six
themes: the first decade of the twentieth century; the transformation of American culture; the automobile and art and literature;
whether the automobile has been beneficial or harmful; the
future; and the historiography of automotive history. Includes
item 636.

632. MacMinn, Strother, "American Automobile Design." *Automobile
and Culture.* Gerald Silk, et al. New York: Harry N.
Abrams, 1984, 209-251.

Traces the history of automobile design through the 1970s.
Focuses on luxury, sport, and experimental cars. Highlights the
role of the custom coachbuilders, who dominated design through
the 1920s. Heavily illustrated.

633. Philip, Cynthia Owen. *Robert Fulton: A Biography.* New York:
Franklin Watts, 1985. xii+371 pp. Bibliography, Index.

Identifies Fulton's genius as the ability to see the
complexity of an idea in all its ramifications and devise an
integrated system. Provides a psychological and not entirely
sympathetic analysis of Fulton.

634. Rae, John B. *The American Automobile Industry.* Boston:
Twayne Publishers, 1984. xii+212 pp. Bibliography,
Index.

Sees the major contributions of the American automobile industry as techniques of mass production and marketing and the organization of large-scale human enterprise. Attributes the relative decline of American companies to the rest of the world catching up. Characterizes government regulation as a "problem" for the industry.

635. St. Clair, David J. "The Motorization and Decline of Urban Public Transit, 1935-1950." *Journal of Economic History,* 41 (1981): 579-600.

Supports the thesis that the decline of urban mass transit was not due primarily to the competition from the automobile, but was a result of a campaign organized by various transportation and energy corporations, including General Motors, Standard Oil of California, and Firestone Tires.

636. Sanford, Charles S. "'Woman's Place' in American Car Culture." *The Automobile and American Culture* (item 631), pp. 532-547.

Assesses the impact of the automobile on the position of women in American society. Concludes that the evidence is inconclusive.

637. Seely, Bruce E. *Building the American Highway System: Engineers as Policy Makers.* Philadelphia: Temple University Press, 1987. xv + 315 pp. Bibliography, Index.

Approaches the engineers of the Bureau of Public Roads as apolitical experts in a democratic society functioning as arbiters of political and financial aspects of American highway policy. Traces the story from 1890 to the rise of the Interstate System in 1956. Focuses on the roles of technical information, cooperation, and consensus in the success of the Bureau of Public Roads. Concludes that the engineers relied on their image of apolitical expertise to influence both general policy objectives and details.

638. Seely, Bruce E. "The Scientific Mystique in Engineering: Highway Research at the Bureau of Public Roads, 1918-1940." *Technology and Culture,* 24 (1984): 798-831.

Reviews the rejection of simple empirical tests for construction materials in favor of a scientific approach to research. Contends that the new approach resulted in difficulties in developing practical knowledge without adding to theoretical understanding; precise data was obtained at the price of increased artificiality. Concludes that no one recognized the sterility of emulating scientists; research had become an end rather than a means to an end.

639. Segal, Howard P. "The Automobile and the Prospect of an American Technological Plateau." *Soundings,* 65 (1982): 78-87.

Reviews American attitudes toward the automobile. Calls for more research into America's characteristic accommodation to technological development.

640. Shaw, Ronald E. "The Canal Era in the Old Northwest." *Transportation in the Early Nation.* Indianapolis: Indiana Historical Society, 1982, pp. 89-112.

Reviews the literature. Argues that the canals were vital to the growth of the region during the years 1820-1840.

641. Stilgoe, John R. *Metropolitan Corridor: Railroads and the American Scene.* New Haven: Yale University Press, 1983. xiii+397 pp. Bibliography, Index.

Explores the reshaping of the American environment by railroads and their right-of-ways from 1880 to 1930. Argues that they nurtured factory complexes and suburbs, but hurt Main Street. Focuses on visual images. Emphasizes the positive aspects of the impact of the railroad.

642. Ward, James A. *Railroads and the Character of America, 1820-1887.* Knoxville: The University of Tennessee Press, 1986. xii+200 pp. Bibliography, Index.

Argues that prior to 1850, the promoters of railroads presented the railroads as a last hope for the unity of the country and the promotion of the general and individual welfare. Finds that railroads were linked to national defense, the alleviation of human misery, and the improvement of national morality. Concludes that by the 1850s, railroad promoters switched from an emphasis of the value of railroads to the nation to the importance of a particular railroad. Draws parallels between the increasing sectional conflict of the 1850s and the competition between railroad companies.

643. White, John H., Jr. "Railroads: Wood to Burn." *Material Culture of the Wooden Age* (item 056), pp. 184-224.

     Demonstrates how extensively the American railroad of the nineteenth century depended upon wood.

644. Yago, Glenn. *The Decline of Transit: Urban Transportation in German and U.S. Cities, 1900-1970.* New York: Cambridge University Press, 1984. ix+293 pp. Bibliography, Index.

     Utilizes quantitative analysis. Rejects the thesis that technological innovations are sufficient to explain the decline in urban transit. Contends that the power and policies of the automobile-rubber-oil industrial complex in the United States resulted in a faster decline of urban transit than in Germany.

645. Zimmer, Donald T. "The Ohio River: Pathway to Settlement." *Transportation in the Early Nation.* Indianapolis: Indiana Historical Society, 1982, pp. 61-88.

     Utilizes the history of Madison, Indiana, to demonstrate the impact of river transportation technology on a community and the decline which resulted in the wake of the railroad.

THE WORKFORCE AND TECHNOLOGY

646. Dubofsky, Melvyn. "Technological Change and American Worker Movements, 1870-1970." *Technology, the Economy, and Society* (item 047), pp. 162-185.

     Summarizes trade union response to technological change as accommodation rather than resistance. Hypothesizes that as workers lost their marketplace bargaining power because automation deskilled jobs, they turned to workplace bargaining power--high-technology, mass-production industries were vulnerable to the shut-down of a single key element in the production process.

647. Gross, Laurence F. "Wool Carding: A Study of Skills and Technology." *Technology and Culture,* 28 (1987): 804-827.

     Distinguishes between understanding the principles of the operation of a piece of machinery and understanding the way in which it was actually operated. Argues that the carders resisted efforts to diminish the value of their skills.

648. Hareven, Tamara K. *Family Time and Industrial Time: The Relationship between the Family and Work in a New England Industrial Community.* New York: Cambridge University Press, 1982. xviii+474 pp. Appendices, Index.

Focuses on the Canadian workers at the Amoskeag Textile Mills in New Hampshire. Combines oral history with printed sources. Argues that the family's influence on the factory system was inversely proportional to the availability of labor. Sees familial interdependence as a response to low economic status.

649. Lankton, Larry D. "The Machine *under* the Garden: Rock Drills Arrive at the Lake Superior Copper Mines, 1868-1883." *Technology and Culture,* 24 (1983): 1-37.

Argues that the drills were accepted by the workforce because they did not lead to the destruction of the social and economic conditions associated with the technology being replaced.

650. Nelson, Daniel. "The American System and the American Worker." *Yankee Enterprise: The Rise of the American System of Manufactures* (item 609), pp. 171-187.

Sees factories in the American System as characterized by technologically defined social relations. Argues that these factories were relatively clean and safe and that they provided workers with opportunities for economic and social mobility.

651. Noble, David F. *Forces of Production: A Social History of Industrial Automation.* New York: Alfred A. Knopf, 1984. xviii+409 pp. Appendices, Index.

Utilizes a study of automatically controlled machine tools' development after World War II to demonstrate that the process of technological development is essentially social, with human thought and action decisive. Rejects the concept of technological determinism. Argues that the choice between numerical control and record-playback was decided in an atmosphere of class conflict and contention over control of the workplace. Provides General Electric as an example.

652. O'Brannon, Patrick W. "Waves of Change: Mechanization in the Pacific Coast Canned-Salmon Industry, 1864-1914." *Technology and Culture,* 28 (1987): 558-577.

Finds that salmon-canning industry mechanized because of the uncertainties in supply, emphasis on speed and efficiency, and desire to replace highly paid skilled workers. Defends the importance of studying microcosms in understanding the macrocosm of technological innovation.

653. Srole, Carole, "'A Blessing to Mankind, and Especially to Womankind': The Typewriter and the Feminization of Clerical Work, Boston, 1860-1920." in Barbara Drygulski Wright, et al, editors, *Women, Work, and Technology: Transformations.* Ann Arbor: University of Michigan Press, 1987, pp. 84-100.

Argues that the typewriter accelerated the feminization of clerical work, but did not provide the initial entry. Finds that women first entered the clerical workforces in large numbers as a result of the Civil War, but remained after the war, predominately as copyists, the least skilled and lowest paid job. Demonstrates that typewriters were marketed as machines for copyists. Identifies the rise of the stenographer-typist around 1880 as the key element in the increase of women in the office workforce.

APPENDIX

Researchers in the history of American science and technology frequently search large numbers of manuscript collections for insights and evidence; two examples pulled randomly from the bookshelf: Robert Bruce (item 016) utilized 137 collections in his study of mid-nineteenth-century American science; David Jeremy (item 588) cited fifty-seven collections in seventeen depositories in his analysis of the diffusion of textile technology. Of course, these numbers tell us only about the successful searches and give no hint about how many inquiries proved fruitless. In some cases, the amount of manuscript material available for a research project is overwhelming. Students of Thomas Edison have over three million pages of manuscripts to study. The Joseph Henry Papers Project has indexed over sixty thousand relevant manuscripts held in nearly three hundred depositories in seventeen countries; another forty thousand Henry manuscripts may not yet have been identified.

Given the huge number of archives and depositories which hold material relevant to the history of American science and technology, some guidance might be helpful to the neophyte scholar. Since providing even a selective bibliography of the guides to such collections is beyond scope of this volume, I am supplying a few thoughts developed during seventeen years at the Henry Papers. Much of what I write will not be new to American historians, but may be of use to historians from other areas of the history of science or technology. I also recommend two recent articles which examine some of the same issues (and others) from the perspective of the archivist: Susan Grigg, "Archival Practice and the Foundations of Historical Method," *Journal of American History,* 78 (1991): 228-239; and Joan Warnow-Blewett, "Documenting Recent Science: Progress and Needs," *Osiris,* 7 (1992): 267-298 (includes lists of disciplinary centers, documentation projects, and catalogs of source material).

There are four major sources for information on manuscript holdings, each of which has its strengths and weaknesses: published material, the *National Inventory of Documentary Sources in the United States* (NIDS), online guides (including the *National Union Catalog of*

Manuscript Collections [NUCMC]), and newsletters. Most major research projects will use most, if not all of them.

My own sense of the current state of affairs is that we are in a transition period. When I started graduate school, the best guides to major collections in the history of American science and technology resided in the minds of the archivists and manuscript librarians. During the past two decades, written guides and databases have become more important, and eventually will replace the archivists or manuscript librarians who know their collections intimately. Yet, at times nothing is as valuable as of correspondence with a knowledgeable archivist or librarian. For instance, in many cases, archivists and librarians alerted us to Henry material in collections where we never would have thought to look.

Published guides are most useful for those who already have some idea of where to look, or have a project which is easily classifiable. While they vary widely in quality, especially in terms of completeness of coverage and indexing, their most serious defect is that they represent a collection or collections at one moment in time. This is a definite problem for students of contemporary science, who need to be aware of the new collections that are being generated continuously. However, even historians of earlier periods have to be careful since material is coming out of private hands all the time, and added to the holdings of major manuscript depositories.

The organizing principle for published guides may be geographical (David R. Larson, editor, *Guide to Manuscripts, Collections and Institutional Records in Ohio* [1974]), thematic (Kathryn Allamong Jacob, editor, *Guide to Research Collections of Former United States Senators, 1789-1982* [1983]), or disciplinary (James R. Fleming, editor, *Guide to Historical Sources in the Atmospheric Sciences: Archives, Manuscripts, and Special Collections in the Washington, D.C. Area* [1989]). Major institutions, such as the American Philosophical Society or Harvard University Archives, have published guides to their collections. There are also highly specialized guides (David K. Van Keuren, *"The Proper Study of Mankind": An Annotated Bibliography of Manuscript Sources in Anthropology and Archeology in the Library of the American Philosophical Society* [1986]).

Not to be overlooked in any search for manuscript material are the footnotes of the available secondary literature. Some historians, the bane of manuscript librarians and their fellow researchers, are inexact or vague in their citations to manuscripts; most, however, are careful and clear. Many provide insights into major collections in bibliographic essays or appendices, and I have noted some of these in my annotations. In such cases, the strengths, limitations, and idiosyncracies of a depository or collection are spelled out. When your guide is a historian of the stature and skill of A. Hunter Dupree (I:556) or Alex Roland (239), the result is a clear sense of the significant collections.

The NIDS is a microfiche publication attempting to provide a "national inventory" of finding aids of major archives and manuscript depositories and a master index to those aids. Finding aids often can give much more useful information than can be found in published guides. By 1991, over twenty-seven thousand aids were available, just a small percentage of the universe of finding aids. The coverage is incomplete and uneven both in terms of the institutions involved and the aids included from any given institution. Moreover, many of these finding aids were created with the assumption that they would be used in conjunction with an archivist or manuscript librarian. A follow-up telephone call or letter is almost mandatory when using the NIDS. Nevertheless, used judiciously with other sources, it is quite useful.

Archivists are confident that the guide of the future will be an online database. Currently, there are two major systems: the Online Computer Library Center (OCLC) and the Research Libraries Information Network (RLIN), which includes NUCMC. There are over three hundred fifty thousand entries in RLIN for manuscript or archival collections, which does not represent the complete universe. Given the cost to depositories of putting data on RLIN, is it not likely to be so in the near future.

One very useful source of information, especially for contemporary science or technology, or in cases where current manuscript holdings are well-known, is the newsletter. Many major manuscript depositories publish newsletters containing announcements of new acquisitions. Examples include the Rockefeller Archive Center, the Library Company of Philadelphia, the National Archives, and the Hagley Museum and Library. There are also a variety of disciplinary newsletters which include similar information: *The Mendel Newsletter*, for the history of genetics; *Chemical Heritage;* the *AIP History Newsletter,* for physics; and the *Newsletter* for the Center for the History of Electrical Engineering. The trick is learning of the existence of the newsletter and getting on the mailing list.

There is one other powerful tool for researchers: the awareness among other scholars that he/she is interested in a particular topic. One of the prized letters at the Henry Papers would have never been located by the Project without such help, because it is a letter from Senator Timothy Howe to his niece, in which the Senator relates a conversation with Henry about Charles Darwin. The letter provides insight into Henry's views of Darwin's theory of evolution. A student at Ohio State, working on an unrelated project, saw the letter, and brought it to the attention of the Henry Papers. As any experienced researcher will tell you, no guide will ever replace serendipity.

AUTHOR INDEX

Numbers refer to entries, not pages

Abir-Am, Pnina, 127
Ainley, Marianne Gosztonyi, 340
Aitken, Hugh G. J., 517
Albury, William R., 366
Aldrich, Michele L., 178, 179, 289
Allen, Garland E. 330, 331, 332, 377
Allison, David Kite, 192, 193
Ammons, Elizabeth, 086
Anderson, H. Allen, 341
Appel, Toby, 361, 364
Armstrong, David A., 170
Arnold, Lois Barber, 106

Baatz, Simon, 128
Bannister, Robert C., 474
Barber, Barnard, 475
Barger, A. Clifford, 362
Barrett, Paul, 620
Basalla, George, 087
Bates, Charles C., 171
Bauer, Henry H., 146
Beardsley, Edward H., 001
Beaver, Donald de B., 015
Becker, John V., 194
Bedini, Silvio A., 595
Belfield, Robert B., 547
Bellomy, Donald C., 398
Benison, Saul, 362
Benson, Charles D., 199

Benson, Keith R., 378, 379, 390
Benson, Maxine, 342
Berkeley, Dorothy Smith, 323
Berkeley, Edmund, 323
Bernstein, Barton J., 195
Bezilla, Michael, 621
Bieder, Robert E., 407, 408
Bierstedt, Robert, 476
Bilstein, Roger E., 196, 483, 484
Bjork, Daniel W., 448
Blight, James G., 449
Block, Robert H., 180
Blouet, Brian W., 446
Blum, Ann Shelby, 088
Blunt, John, 277
Bodansky, Joel N., 036
Borell, Merriley, 363
Boydston, Jeanne, 534
Boyer, Paul S., 613
Brobeck, John R., 364
Brodhead, Michael J., 344
Bromberg, Joan Lisa, 197
Brožek, Josef, 450, 451
Bruce, Robert V., 016
Bruchey, Stuart, 047
Bruins, Derk, 045
Buck, Peter S., 107, 198
Buckley, Kerry W., 452, 453
Bud, Robert, 276
Bulmer, Martin, 399, 477

SUBJECT INDEX

Numbers refer to entries, not pages